Rewilding

Rewilding

India's Experiments in Saving Nature

Bahar Dutt

OXFORD
UNIVERSITY PRESS

OXFORD
UNIVERSITY PRESS

Oxford University Press is a department of the University of Oxford.
It furthers the University's objective of excellence in research, scholarship,
and education by publishing worldwide. Oxford is a registered trademark of
Oxford University Press in the UK and in certain other countries.

Published in India by
Oxford University Press
22 Workspace, 2nd Floor, 1/22 Asaf Ali Road, New Delhi 110 002

ISBN-13 (print edition): 978-0-19-947411-0
ISBN-10 (print edition): 0-19-947411-7

ISBN-13 (eBook): 978-0-19-909833-0
ISBN-10 (eBook): 0-19-909833-6

Typeset in Perpetua Std 13/17
by Tranistics Data Technologies, Kolkata 700 091
Printed in India by Replika Press Pvt. Ltd

To Vijay
for keeping the laptop (and my soul) charged on many a
rewilding journey

The mammoth and the dodo never saw it coming—
in the end, there is only the idea of species, like a chair
left swinging when the kids go in for lunch....

—Andrew Slattery, Longing

Contents

Contents

Preface

In my mind's eye, I had been scouring the world, seeking out all that was wrong and needed to be fixed. I saw air laden with carcinogenic particles, plastics choking marine life, and sludge pouring into weary rivers. My soul was weary too. I had become too used to seeing the world consistently deprived of wilderness, carpeted with concrete. My heart ached for a rich forest where species of all shapes and forms thrived.

I stared out of my window one languid winter afternoon, my cup of coffee as bitter as my outlook. It was that time of the day when my toddler was asleep and I took a pause between the whirlwind morning behind me and the evening that lay ahead. The mid-afternoon reverie was broken by the sight of a coppersmith excavating a nest for itself; a pair of grey hornbills chased each other through the maze of overhanging branches that bent over our house, even as the shoots of new leaves pushed themselves out from the old trees that lined the main road. Even this mosaic of urban nature that stared at me through my study was always in a state of flux, constantly fighting the onslaught of the human footprint. The ladybirds would soon arrive in the botoxed park behind our house, offering a glimpse of their tomato-red wings to my toddler before vanishing. Tiny spiders hard at work clinging on to their silken nests would be swept away by the overzealous gardener the next day. Nature was constantly attempting to push its shoots out, to renew itself, wherever it could. And in that instant, I realized it was time to renew my interpretation of nature as well.

After a decade of covering stories of poaching and deforestation, battling the mining mafia, and exposing corrupt politicians, it was time to tell the stories of efforts being made to protect nature. As I viewed nature through this new lens, I found many such instances—of turtle eggs incubating in riverside nurseries and then released in cool waters, rhinos abounding in areas from where they had been wiped out, tigers colonizing new habitats. India was on the move, as was its wildlife. But my research had a greater mission than documenting these feel-good stories. I wanted to explore a concept called 'rewilding' that had become a hot topic in the

West. The term, first used by activist Dave Foreman (1972) and later refined by biologists Michael Soulé and Reed Noss, was revitalizing the conservation scene across the world. Rewilding seemed more appealing than traditional conservation, as it gave humans a mandate to fix the problem. I wondered if the term could be applied to India.

And so, I dedicated two years of my life to travelling around India in search of projects that could fit this definition. At each project site I visited, I bore witness to some of nature's rare phenomena—the murmuration of starlings above a recently restored grassland habitat; the flapping of the mahseer fish as it jumped back into the pond where it was being bred in captivity; the shy, hunched-over vulture peering out through the bars of a cage; and the delightful antics of the smallest pig in the world as it sniffed for earthworms, the sun dancing off its bristly back. Nature was being restored in different ways across India: by captive breeding, by releasing species into the wild, or by revamping habitats to support more diversity.

My mission, as already outlined, was to investigate the practical application of efforts to restore and renew nature. While some of the projects that are discussed in this book may be familiar, what remains untold are the sub-plots, the many twists and turns of a journey to heal ecosystems, reintroduce species, or remove alien weeds and replace them with grasslands. I was keen to analyse if the effort to repair nature offered real-life lessons in conservation that could be replicated in other parts of India. But this book ended up doing so much more. It presented an opportunity to encapsulate the story of individuals fighting poaching, terror, and even disease outbreak to beat the challenges before them, to fight for the planet, species, and biodiversity.

This isn't a book that just celebrates green heroes; rather, it is one that notes the tremulous journey of conservation and how easily it can suffer a setback, often from situations outside the control of natural science. Despite the setbacks, what we need to remember is that the experiments documented in this book may be limited in time and scale, but are still a much-needed salve on the open wounds we have inflicted on our natural world.

Bahar Dutt
August 2019

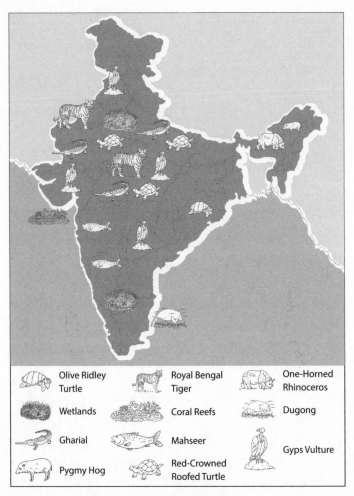

Species and Habitats Featured in This Book

Source: Author.

Note: This map does not represent the authentic national and international boundaries of India. This map is not to scale and is provided for illustrative purposes only.

1
Introduction

Humans have been restoring nature long before forest officers, conservation biologists, and tree huggers arrived to take up the fight. Local communities who lived close to forests set aside land as sacred groves and imposed no-hunting seasons in order to give wild species and nature time to renew and repair. Modern-day restoration of ecology may, however, have risen as a discipline to undo all the damage we have inflicted on the natural world. India has a network of over 500 national parks and sanctuaries, and over three decades of environment laws in place to protect its rich natural heritage, along with a battery of ambitious international agreements to affirm its commitment to protecting nature. Yet the environmental crises of our times cannot be ignored, with roads and railway lines cutting through forests, more and more species featuring on the endangered list, and grasslands being wiped off like grease from a dinner plate.

The concept of rewilding creates hope; it gives the opportunity to set things right, to link protected areas through corridors, to bring back species once lost, and to revitalize our forests, rivers, and wetlands with all forms of life. While it was first coined by Dave Foreman on aesthetic and moral terms (as our ethical responsibility to protect other species), it was really biologists Michael Soulé and Reed Noss (in a landmark paper, titled 'Rewilding and Biodiversity: Complementary Goals for Continental Conservation', published in 1998) who gave the term a scientific basis. According to the authors, rewilding is a conservation method based on three Cs: Cores, Corridors, and Carnivores. Subsequently, the definition was expanded to include keystone species—creatures that interact with the environment so strongly that their absence would significantly alter the habitat. A classic example of keystone species are elephants (they push over trees and keep vast grasslands in the Serengeti intact) or beavers (they work as ecosystem engineers by transforming a river into a pond or a swamp).

Put simply, rewilding gives us the opportunity to kick-start and reboot, to restore natural processes and set right the wrongs that we have committed. Rewilding emphasizes that it is no longer sufficient to declare an ecosystem or a landscape as protected; it must be restored with species and corridors. Of course, this is not to say that before the term became popular there were no efforts by humans to save habitats or species. While the term is new, the concept is not. 'Restoration ecology' or 'reintroduction' belong to a field of conservation science that aims at renewing and restoring degraded, damaged, or destroyed ecosystems and habitats in the environment by active human intervention and action. What rewilding does is fix the jigsaw puzzle by reconnecting corridors or by placing back the 'missing' species that had vanished from an ecosystem.

Countries across the world have embraced this concept; they aim to restore large-scale wilderness areas or corridors between them, or introduce apex predators or keystone species. These projects include the Yellowstone to Yukon Conservation Initiative in North America (also known as Y2Y); transboundary projects, including the Peace Parks in Africa; community-conservation projects, such as the wildlife conservancies of Namibia and Kenya; or those in Europe that involve the reintroduction of species such as the Iberian lynx. In 2011, the Rewilding Europe initiative was established with the aim of rewilding 1 million hectares of land across the Iberian Peninsula, from the Carpathians to the Danube

Delta, by 2020 'for the benefit of nature and people'. The idea is to reintroduce species that are still present (in isolated pockets or smaller numbers) in Europe, such as the Iberian lynx, Eurasian lynx, wolf, European jackal, brown bear, European bison, red deer, griffon vulture, black vulture, Egyptian vulture, and the Great white pelican. The goals include restoring grasslands and ecosystem functions (including predation), and creating a rural tourist economy. In the UK, the non-profit Rewilding Britain initiative, launched after being inspired by journalist George Monbiot's seminal book on the subject, plans to bring back such species as the lynx, beaver, and wolf. Rewilding projects may require ecological restoration or wilderness engineering, particularly to restore connectivity between fragmented protected areas, and reintroduction of predators where they've been wiped out.

Different countries have borrowed the concept and created their own version of it, which begs the question—can the concept be applied to India, where biodiversity and humans live together like tossed salad, where protected areas don't exist on the same scale as in Africa or Europe? This was a particular concern, because the idea, as envisaged by Soulé, was to focus on the restoration of big wilderness areas based on the regulatory role of large predators. In India, on the other hand, land that is devoid of humans for the introduction of large predators is limited in supply. And can rewilding really be the panacea to all modern conservation problems? These were some of the questions I sought answers to as I set off on my journey across the country.

Indians were moving animals around as a conservation tool long before the term 'rewilding' was coined, or even before tranquillizer guns and sophisticated aircraft arrived to enable this process. The first recorded instance could possibly have been in 1928, when the ruler of Dungarpur (today a city in South Rajasthan), Maharawal Lakshman Singh, became concerned that not a single tiger was to be found in the dense forests of his principality. All the tigers, it seemed, had fallen prey to *shikar* or hunting, which was very popular in those days. At the time, the maharawal (the title that got gradually deflated to the title of maharaja or king) was still a minor, so his principality was run by a British political agent, Donald Field. Field had a fondness for shikar, and it seemed that he and his friends had managed to wipe out Dungarpur's tiger population. The furious maharawal shot off a letter to Delhi, demanding an explanation. The British rulers, to humour the maharawal, suggested that tigers be brought in from Gwalior. And thus took place the country's first initiative to translocate tigers.

Two years later, in 1930, a male and a female tiger were caught, caged, and put on a train from Gwalior to Talod in Gujarat. But Talod was still some 80 kilometres away from Dungarpur. It's a miracle that the animals survived the journey, which was mostly on bullock carts. Once the tigers reached his kingdom, the maharawal banned poaching. He even stopped his brothers from hunting in the forests—as a result, the tiger population steadily increased. At the time of Independence, 25 tigers were reported in the forests of Dungarpur. In the years to come, Dungarpur was once again tragically emptied of all its tigers because of poaching.

Post Independence, in 1956, the Indian Board for Wildlife accepted a proposal by the government of Uttar Pradesh to establish a sanctuary for the Asiatic lion in the Chandra Prabha Wildlife Sanctuary, covering 96 square kilometres in eastern UP, where climate, terrain, and vegetation are similar to conditions in Gir, in Gujarat. In 1957, one male and two female wild-caught Asiatic lions were set free in the sanctuary, and in 1965, 11 animals were recorded, but which disappeared thereafter. These, though, were the early days of reintroduction, when not much attention was paid to the condition of the habitat, the dialogue with local communities, or perhaps the threat from poachers. The project was obviously short-lived. In 1984, a plan was made to translocate the one-horned rhinoceros from Pobitara in Assam to be released in UP's Dudhwa National Park, where the species had once roamed freely. Nine rhinos were brought in from Assam and Nepal, and released in an area surrounded by an electrified fence over two years; of these, seven survived. Over the years, this population multiplied several times and currently stands at over 30 individuals. However, the programme had limited success. Given the small size of the founder population and the fact that they were confined within an enclosure, this could never really be considered rewilding in its truest sense.

So can any release of animals in an ecosystem be considered an example of rewilding? The answer is, of course, an emphatic no. Rewilding is a far more complex process. As I set out in search of more contemporary case studies that fit the lens of rewilding, it was obvious that as a Western theory, with its focus on restoring large swathes of land, or bringing back charismatic megafauna to an ecosystem, rewilding has limited applicability in India where wildlife exists in a mosaic of human-dominated landscapes. The scale of rewilding in India is incomparable to efforts in the Russian tundra, the South African reserves, or North America, where rewilding has become an alluring term. And so here I must issue an alert: I have used the term with a rider. India is not home to a plethora of rewilding projects as conceived by Western academic literature or on a scale comparable to societies where human population densities are low. Given the current environmental crisis, especially the siphoning of forests to feed a growing economy, the space for any rewilding to occur is, in fact, only shrinking. Further, of the three Cs that Soulé identifies, the projects discussed in this book probably address perhaps just one or two.

For me, rewilding, as conceptualized by George Monbiot in his seminal 2013 book *Feral: Rewilding the Land, the Sea, and Human Life*, holds greater appeal. For Monbiot, the term can have a number of definitions. First, it is the restoration of environments to their naturally occurring states. Second, it returns the thrill of nature to human existence—something that is gravely missing in the technological age. But most importantly, rewilding puts the natural world into focus on a number of levels. As a storyteller of conservation, to me, it creates hope and restores faith in the ability of humanity to rebuild ecosystem processes and restore nature.

In selecting examples of rewilding to be analysed in detail for this book, I chose three criteria. The initiatives had to focus on a species rather than on just a few individuals (conservation biology by its very definition focuses on saving species and not individuals); they would have to represent a diversity of ecosystems, species, and landscapes; and lastly, the effort must be committed to the goals of conservation—in other words, the effort must not stop just at captive breeding, but also have a strong link to restoring ecosystem processes or improving the habitat where the species was to be released. That's why, while many well-meaning projects involve planting trees on a massive scale, or releasing individual injured animals back in the wild, they were not considered as being within the ambit of this book. Animal welfare, for instance, looks at healing or looking after injured animals irrespective of their status in the wild; conservation focuses on restoring species as a whole.

The case studies described in this book are distinct from the simple translocations done in the 1970s and 1980s in India: here, the focus is not just on shifting animals, but also in improving their habitat or undoing the wrongs of the past. For instance, efforts to bring back vultures to our skies isn't just about breeding the birds in captivity but also about trying to eliminate the toxins that nearly wiped them out from their natural habitat. That a developing country took on this challenge even when rich countries in Europe have still not banned the drug is significant.

In recent times, one of the most iconic visuals of restoring species occurred in 2008. On a rainy morning

9

in June that year, an Indian Air Force helicopter took off with a consignment it may have never carried before when two full-grown tigers were airlifted from one part of Rajasthan to another. The journey these big cats undertook was perhaps less arduous than the one their ancestors undertook in 1930 to Dungarpur, courtesy the right dose of tranquillizers and the swift action of the choppers that airlifted them. The tigers were on their way to Sariska Tiger Reserve, which had lost its entire population to poaching. (Of course, this effort too was fraught with problems; two years later, one of these tigers was found dead, with the authorities clueless about the cause.) Panna, which had also lost all its tigers to poaching in 2009, made a similar effort to translocate tigers, and turned its story around by clamping down on poachers and building bridges with the hunting community that was once active around the park. From a founder population of three, Panna's tigers at last count stand at 11 adults and 14 cubs.

These attempts seemed to fit in well as examples of rewilding, fulfilling at least two of the three Cs that Soulé outlined—carnivores and cores. The concept of corridors is tricky, since these do not have legal protection in India.

The effort to revive populations of the pygmy hog, through a captive-breeding programme and then their subsequent release in the wild through restoring grasslands, is another classic example of rewilding. Backed by some of the best scientific minds in conservation, the pygmy hog centre in the Northeast has managed to stabilize populations of this critically endangered pig, and has been discussed in the book. The project stands out not just for its efforts to conserve a species, but to bring the focus back on a type of ecosystem that is itself threatened, the grasslands. In India, grasslands face a high degree of vulnerability due to anthropogenic as well as grazing pressures, habitat fragmentation, and proliferation of invasive species. In fact, some of the most threatened species depend on grasslands as an ecosystem for their survival, such as the Hispid hare, Bengal florican, and the almost-extinct Great Indian bustard.

Another grassland species, the Indian rhino, in spite of the Dudhwa setback, was reintroduced in protected areas in Assam, in particular Manas National Park that lost most of its wild species to years of insurgency. Once reduced to nil, the rhino population in Manas today stands at 33. The park has been on the rebound since a political ceasefire was established with the insurgents. Other species like the swamp deer have also been brought into Manas. Rewilding the park has not only revived its glory but also increased tourism inflow. Efforts are now on to create a Tranboundary Manas Conservation Area

(TraMCA) as a larger space that connects protected areas, biological corridors, and the adjoining reserve forests of southeastern Bhutan with that of northeastern India. If this materializes, it could address Soulé's third 'C'—corridors.

The rewilding projects in this book have focused not just on terrestrial species; some small-scale interventions that impact marine ecology have also been discussed. India has a coastline that stretches for 7,000 square kilometres, with over 31 protected marine areas and over 100 protected areas in the archipelago of the Andaman and Nicobar Islands. Some sporadic efforts have been made to rewild the seas by planting seagrass in degraded areas off the coast of Tamil Nadu or through the regeneration of coral reefs off the coast of Gujarat; these have obviously not taken place on the large scale that Soulé believes is needed to be called a rewilding effort. But as small and isolated as these examples may be, they are relevant given the existing threats to coral reefs from rising temperatures and the crises facing marine life from plastic pollution. The extinction of coral reefs would potentially lead to a loss of global biodiversity at a catastrophic scale. The bleaching of corals due to rising sea temperatures is expected to cause further damage to marine ecosystems already suffering from over-extraction and climate change. Can the endeavour to artificially breed corals also help local fishing communities? I examine the intricacies of rewilding our seas in the book.

Perhaps the most fascinating example of rewilding that I came across was the one focused on the mahseer. The effort by a private company to breed an endangered fish in captivity may seem well meaning, but what happens when they are released in the wild? Another reason I selected this project is because it moves away from the charismatic mammals that are generally the targets of high-profile, glamorous initiatives.

Rewilding efforts need not be focused just on protected areas in far-flung locations or with elephants and tigers; they could be about bringing back pollinators to our cities, reviving a wetland, or reintroducing a bird species that once frequented a neighbourhood. It is for this reason that I've profiled two urban rewilding projects— one from Gurugram and the other from Bengaluru. Both these case studies highlight how urban citizens did more than just plant trees; they succeeded in restoring the ecosystem functions of highly degraded landscapes. In Gurugram, the urban rewilding project succeeded in healing nearly 550 acres of land and turned it into a biodiversity reserve. Today, the Aravali Biodiversity Park functions as a repository for the flora and fauna of the Aravali mountains; it has a water conservation zone, as well as an educational space to spread awareness about environmental issues. While their efforts have yielded good results, the question to ask is whether such efforts lead to the exclusion of the urban poor communities who may have once used these spaces as a commons. A harder question: are urban rewilding projects worth it? For instance, some scientists have argued that it is better to invest money in protected areas that are home to a higher percentage of wildlife than in urban restoration efforts. These are the issues discussed in this chapter.

As I set out on my journey documenting the case studies in this book, interacting with forest officers, local communities (who live around national parks), and scientists, the first revelation I had was that there was so much good work happening around the country. Some, like the vulture breeding programme, were well known, but the contours of the programme, the obstacles along the way, and the daily victories were not. In the details of each of these case studies lay potential lessons for other rewilding projects that scientists may wish to take up in other parts of India.

I would like to emphasize, however, that I do not, through this book, intend to create the image that India is a hotbed for rewilding projects, especially given the scale at which the theory has been practised in other countries. What I have tried to outline are some attempts made on restoring nature that go beyond conventional conservation practice.

Rewilding is a recent term. While many scientific papers have been written about it, the books that deal with the subject are few. Dave Foreman wrote a book titled *Rewilding North America: A Vision for Conservation in the 21st Century*, in 2004. Foreman describes recent discoveries in conservation biology that call for networks of wild lands instead of isolated protected areas, and, reviewing the history of protected areas, shows how such networks are a logical next step for the conservation movement. The final section describes specific approaches for designing such networks, which is based on the work of the Wild Lands Project, an organization he helped found, and offers concrete and workable reforms for establishing them. The problem with such an approach is that it assumes there are vast vistas of land that are free of humans available for rewilding. Monbiot, while acknowledging the rights of communities who live off the land, misses the sheer impact of densities of anthropogenic pressures. He takes readers on a journey around the world to explore ecosystems that have been 'rewilded'; essentially, they have been freed from human intervention and allowed—in some cases for the first time in millennia—to resume their natural ecological processes. For this, he 'kayaks among dolphins and seabirds off the coast of Wales and wanders the forests of Eastern Europe, where lynx and wolf packs are reclaiming their ancient hunting grounds'. Both these books promise an Eden that may be easy to implement in the Western world, where a high density of human populations is not of concern. Large tracts of land that can be converted to areas for biodiversity is easy to aim for and is an appealing concept, given the fewer contestants for land in Europe, the UK, or the US. The question then is, how do we build the case for rewilding in other parts of the world where biodiversity thrives in a sea of humanity?

And should one conclude then that rewilding is a largely Western import? Caroline Fraser draws the bridge between the East and the West through her book, *Rewilding the World: Dispatches from the Conservation Revolution*, which offers examples from the Americas to Africa and Asia. Her book claims to offer the first definitive account of a visionary crusade to confront the extinction crisis and takes examples that are set not just in the developed but also the developing world. This book takes us around the world, looking at efforts to rebuild wild ecosystems and give species the habitats they need to survive while addressing the politics of conservation, the human factor, and the role of communities in this process. Fraser cites leading scientists and environmentalists to explain the cutting-edge science and political action that has begun to rewild parts of the Earth and help to rebuild the environmental services that sustain us. She takes examples from Nepal, Kenya, South Africa, and many countries that have worked with communities to achieve the goals of rewilding and is thus able to bridge the North–South divide on rewilding better.

North or South, it has to be remembered that rewilding cannot be the panacea to all our conservation problems. The case studies documented in the following chapters highlight achievements, but they also focus on how perilous each victory is. No sooner do you celebrate the release of a rhino in a park that news can come in of its poaching. Rewilding may sound like an exciting proposition, but it is tough, complicated, and requires political and financial commitment if we indeed want to fulfil the purpose of conservation. And there are problems galore. The worst is the die-off of species once released, or other factors impacting the success of the project that are beyond the control of scientists.

The proposed restoration of ecosystems through the reintroduction of species is seen by many as a way to stem the loss of biodiversity and the functions and services that it provides to humanity. In spite of the complications involved, rewilding offers nature and humans a second chance. In addition, there is the hope that it might lead to increased public engagement and enthusiasm for biodiversity. The purpose of this book has been to present examples of introductions or reintroductions— those that have succeeded and those that have failed, provoking unexpected negative consequences. But the aim is to kick-start a new line of thinking so as to protect biodiversity, and to hold on to the strand of conservation optimism it creates.

Lastly, while species and habitats must be protected *being qua being*, let's not forget that conservation of forests and diversity is beneficial to humans. With all the myriad problems that conservation throws up, it offers ecosystem services like fresh water, fertile soil, clean air, and protection against flooding or extreme weather. The benefits from conserving nature, according to one estimate conducted by Robert Constanza and his colleagues in 1997, was calculated to be $33 trillion (equivalent to $44 trillion at today's rates) if given a value, one that we often tend to ignore.

Rewilding as a conservation tool has endless possibilities: we could use it to revitalize the mosaic of wetlands that have been converted into garbage dumps, create corridors of forests that benefit communities and wildlife, or even bring back the house sparrow to our neighbourhoods. Of course, we must remind ourselves that rewilding is not as simple as opening the cages in a zoo to let animals roam free, nor is it a modern-day version of recreating wildlife parks. A rewilding effort cannot create an unreal or an unequal world; it requires huge financial commitments, and as the proceeding chapters will show, is a constant work in progress. The conservation Eden may never be attained, but at least rewilding shows us the yellow brick road leading to it.

2
Bringing Back Stripey

◎

COMMON NAME: Bengal Tiger [English], Bagh [Hindi]
SCIENTIFIC NAME: *Panthera tigris tigris*
CONSERVATION STATUS: Endangered

◎

I am bundled in an open jeep, nearly dozing off, when a cold, early-morning burst of wind slaps me into wide-eyed wakefulness. I sit up and strain my eyes to catch a glimpse of the yellow stripes that could send any slender-legged ruminant running for cover. This is undulating tiger country: scraggly trees holding on to large grey boulders with their exposed roots, giving way to the gushing Ken River. The terrain here, in the Vindhya mountains in northern MP, consists of extensive plateaus and gorges. This area is the northernmost tip of natural teak forests and the easternmost tip of natural kardhai (*Anogeissus pendula*) forests.

We have been driving for over an hour. The lemon rays of the sun have only now penetrated the canopy, opening up a kaleidoscope of brown and green on the forest floor. I am unable to spot a tiger, but I see a grey jeep parked in the middle of the dusty yellow tracks of the tiger reserve. The park guide accompanying me shakes his head apologetically when I request him to pull up next to the vehicle. 'They are tracking the radio-collared tiger. We are not allowed to disturb them,' he explains. We are, after all, in one of the most controversial tiger reserves in the country.

As the forest track narrows down, trees bend over as if joining hands on either side. A sambar deer looks at us quizzically, pauses for a moment, and then melts into the foliage behind. (Apart from the tiger, Panna is home to other animals like the leopard, nilgai, chinkara, chousingha, chital, rusty spotted cat, porcupine, and sambar.) The high point of the safari is the drive up to a plateau known as Vulture Point, where we stop for breakfast. Down below is the Dhundwa Seha, a deep gorge that is supposed to be an ideal nesting place for vultures. Seven species have been recorded in the park—Long-billed, Red-headed, Egyptian, and White backed, while the migratory species include Eurasian griffon, Himalayan griffon, and Cinerous vultures. Today, there is much excitement: a 100 feet below, a leopard slinks through the trees, sending tourists into a frenzy.

All tourist jeeps entering Panna Tiger Reserve stop at Vulture Point for sweeping views of the plateau and the dramatic gorge below—and perchance the sighting a tiger? Today it's the leopard that's caught everyone's attention. Once that excitement has settled down, we unwrap our sandwiches and sit on a ledge, letting hot coffee pervade our system and warm our souls.

Later, when they are off-duty, I get the chance to talk to the forest staff we had encountered inside, a team that tracks Panna's most precious residents through the *tic-tic-tic* of a radio collar. Twenty-eight-year-old Girwar Singh is a forest guard with the MP forest department. 'The tigress was showing her young cubs how to hunt a chital. They were running everywhere; it was tough keeping track of them,' he tells me. Singh is just one of the army of foot soldiers who is following the radio-collared tigers night and day through Panna's deep valleys and gorges. He is in charge of two tigers, T-1 and P-141. The periodic *tic-tic* of the radio collar sending out a signal to the antennae he holds above his head from time to time lets Singh know that the tigers are safe. This young man has a ringside view of tiger conservation in the park, tracking the big cat and learning more about its life. Such knowledge, though, comes with hard work: Singh's duties entail physical and mental stress. What causes the stress? He cannot lose track of his animals—ever. Sometimes there is panic; if he can't hear the sound, it means either the tiger has wandered off into a valley, or the radio collar is not working. He must then alert senior officials and an arduous effort is launched to track down the animal.

Tracking tigers on foot is not only tedious, it's also expensive. The state government has spent millions of rupees to turn the story of Panna around—developing surveillance teams that monitor the big cats, training their forest guards, even relocating villages from the core area. In 2009, Panna made international headlines. Like Sariska Tiger Reserve in Rajasthan, it lost every single tiger it had to poaching. The sordid tale of the disappearance of Panna's tigers may not even have reached a wider audience if it hadn't been for Raghunandan Chundawat. Raghu had been researching the big cats in this part of Bundelkhand for over a decade, living deep inside the national park, following the radio-collared tigers day and night. It was he who shot off letters to the National Tiger Authority and to the ministry of Environment and Forests to let them know that Panna was going the Sariska way: soon there would be no tigers here. It took four years for the government to act. A special investigation team constituted by the MP government in 2009 confirmed Raghu's worst fears—it could not find evidence of even one tiger. Raghu should have been the hero of Panna; he became its fall guy. His research permits were cancelled and the state hounded him out of the tiger reserve.

Once known for its diamond extractive industry, Panna was the jewel in the crown for the state as far as tiger conservation was concerned. In its heydays, there were more than 35 tigers here. After the big poaching debacle, Panna had to sketch a new story for itself, one that involved bringing tigers from other national parks. Now, almost 10 years since the sun came down on its tigers, the story of Panna has taken a positive turn. An embarrassed forest department, flummoxed by the complete wipeout, rushed to restock the forest, bringing in tigers from nearby reserves like Kanha and Bandhavgarh. Today, Panna has nearly 11 adult tigers and 14 cubs as per the last estimates conducted by Wildlife Institute of India (WII) in 2013. Panna is on the rebound, but are the reintroduced tigers safe? And what are the new emerging threats? To answer these and many more questions, I made my way here in the spring of 2017.

Many large carnivores are currently in danger, with around 61 per cent of them designated as 'threatened' by the IUCN [International Union of Conservation of Nature] Red List of Threatened Species. Most large-carnivore reintroductions have been done in North America, Europe, and South Africa, where declines have ensued from the animals' direct persecution by humans. 'Once the factors responsible for the original extirpation of a large carnivore have been removed, reintroduction has proved a viable conservation option given the backdrop of low human densities, extensive land availability and the commitment of adequate financial and socio-political support for the reintroduction project,' argue M.D. Madhusudhan and A.J.T. John Singh in a 2009 paper titled 'Tiger Reintroduction in India: Conservation Tool or Costly Dream?' The reintroduction and recovery of the Florida panther (*Puma concolor coryi*) in Florida, USA, during 1981 and the reintroduction of the African wild dog (*Lycaon pictus*) in Africa in the 1990s are two such instances, among others, that have enriched the science and management of carnivore reintroductions. In India, the motivation to reintroduce tigers has more often been political than scientific, embarrassed as state governments have been by the negative media publicity it has brought them, and, of course, the loss of tourism revenues with the disappearance of the top predator from their reserves. Nonetheless, both Sariska in Rajasthan and Panna in MP have been working hard to keep their tigers safe, though in the former some incidents of poaching have been reported.

At the edge of Panna, close to Ken River, the grass glistens in the day, sending out soft, murmuring whispers. The grass is so tall, it could camouflage a big tiger. Here, Raghu has started a new innings. Along with his wife Joanna Van Gruisen, a wildlife photographer and writer, he's opened a resort for tourists visiting Panna. Hounded by the system, with his research permits cancelled, Raghu is now a quiet man, but not defeated. He's sitting in the open courtyard of his resort, hammering away at his laptop; he wants to tell the story of Panna through a book that is soon to be published. It's quite a career switch for a scientist who once taught at the premier WII, but Raghu seems to have adjusted to his new avatar. Summer is just setting in, and even in March temperatures are about to touch 40°C. I am offered an ice-cold glass of lemonade with a hint of mint, to recover from my long journey here.

'We will sit in the evening,' says Raghu; I nod, taking refuge in the cool interiors of my mud-plastered room.

As night descends, a lazy toad hops across the stone pathway that's retained the sun's heat. The gentle flicker of lamps guides me to the main hall, where tourists gather for the evening. It's mostly foreign tourists on their way to Khajuraho who make a stop at Panna. There's a bird enthusiast from Surrey, UK, who is hoping to spot a four-horned antelope, locally known as the chousingha, on his safari the next day. Raghu listens patiently as the birder chatters incessantly about his encounters in the wild. The rest of the guests stare out into the dead night, taking in the view of silver stars against a dark sky.

Once Raghu has finished attending to his guests, I bring up Panna's turnaround story. As a scientist, is he convinced that the park is now a safe haven for tigers, that mistakes of the past have been addressed? Is he happy to be back since the debacle of 2009? Raghu is clear: 'I didn't come back to keep an eye on the forest department. If they ask me for advice, I am here, but that's not my prime motivation for being in Panna. My problem with the department is there is no vision—for tiger conservation or even management of wildlife,' he explains, frustrated. But years of dealing with the department has taught him to leave his bitterness behind. Raghu has plans beyond the park; he wants to focus his research on the less glamorous species like the Indian fox, the hyena, and the grey wolf. He's keen to get a large student community assisting him in this research. While Raghu may have reconciled himself to this new role, I find it hard to believe he will give up his passion for tigers.

The next day, I have a meeting with Vivek Jain, the field director of the park, at his office in the town of Panna. The field director is sitting in a black swivel chair and seems affable, open to talking. Unlike most forest officers who seem nervous about talking to the media, Jain is calm and answers most of my questions with a smile. He starts by saying, 'I have been here only 14 months. A lot has happened before me. Our job is now more difficult, since there are so many tigers, so monitoring them becomes a challenge.' He turns on his desktop to show me a map of the reserve that's full of green lines zigzagging across it. These lines represent the movement of the radio-collared tigers. The map dispels all those myths of tigers who live only in forests. Tigers don't know boundaries; they move through forests, rivers, and agricultural ecosystems.

Jain believes it is only male tigers that stray out. When I ask Raghu this question later in the evening, he counters this claim. 'Even female tigers move,' he explains, adding, 'though perhaps not over such long distances as their male counterparts.'

Those zigzagging lines on the map on Jain's computer clearly and visibly show tigers moving out of the park. That's why some of the best tiger experts in the country are emphasizing the need to conserve habitats beyond national park boundaries. Animals need to disperse; they don't just live in forests. A male tiger may disperse as far as 200 kilometres from where it was born—so tiger experts are reorienting management decisions towards this reality. I'm not sure if Panna is following suit. I ask Jain what he does when a tiger goes beyond his park boundaries, and if anyone is watching out for it. 'Is it a Panna tiger or a tiger on its way to Kanha? Who is responsible for that individual?' I ask in jest.

Jain retorts, 'It is the responsibility of the forest officer in that division to look after this tiger.' Pointing to his computer screen, he says, 'See, for instance, right now this tiger is in a "territorial forest". [Forests are divided into territorial and wildlife divisions for management purposes.] The staff over there does not have the facilities to follow this radio-collared tiger or have a tranquillizer gun. In such a case, we have to provide assistance. In another instance, as with T1 since he was the founder tiger [the tiger brought in from another park to restock Panna], we tranquillized him and brought him back several times.' Then, perhaps tired with my barrage of questions, he says, 'You should meet our local veterinarian Sanjay Gupta.'

The next day finds me back at the forest office, this time to meet wildlife veterinarian Sanjay Gupta. Surrounded by bottles and syringes, Gupta is busy planning a 'rejuvenation camp' for the captive elephants of the forest department. 'We give them a lot of rest. They are massaged and put on a special diet with vitamin and mineral mixtures,' he explains proudly. It is his task to manage the tigers when they stray out. Gupta, in his 17 years at the park, has the unique record of tranquillizing 55 tigers and bringing them 'home'. 'I was here when there were 32 tigers. I was here when the number went down to zero. Now, when there are estimated to be more than 40 tigers, I'm still here,' he tells me.

The good doctor arrived in Panna fresh out of veterinary school, an appointment that, in government parlance, is referred to as a 'punishment posting'. It was absolute wilderness, cut off from any big town. Now, years later, this unassuming doctor could well be the hero of the story. Apart from the 55 tigers that he has tranquillized, he has on his list hyenas, elephants, and several other wild species. He describes every rescue effort with panache, making it all look so easy. This is the stuff that wildlife documentaries are made of. Unlike the intrepid scientists shown as heroes in these documentaries, this man's work may never be celebrated. Gupta shows me videos of him darting tigers on elephant back, of releasing the cats, of cleaning the wounds of injured tigers. And it's all in a day's work, with no fancy cameras following him. They are edited like cheap wedding videos, but the content is what wildlife filmmakers yearn to capture in their films.

Gupta tells me that he's caught and vaccinated more than 3,000 dogs around the park. This is to ensure that no disease like rabies is passed on to the big cat, especially since that incident in 2013 of a tiger cub being bitten by a stray dog.

His most challenging assignment was a few years ago, when one of the tigers strayed out of Panna into a village nearby. She was finally located in a paddy field flooded with water, making rescue operations formidable. Describing the incident, Gupta says, 'We had to tranquillize the tiger in an open field that had many wells—and it was raining. She could have fallen into any of the wells during the darting exercise. So we surrounded the tiger with our captive elephants, and then I darted her with my tranquillizer gun. By this time, hundreds of people had gathered. We had to call the police in for protection. What's more, this happened during Durga Puja. People thought the tiger's presence was auspicious, so they started throwing coins on her even as we put her on a stretcher and carried her out on a tractor. Imagine if, in this sea of people, the tiger had woken up and attacked someone.'

I leave him there in his room, preparing a tranquillizer gun for a run into the national park, thinking about what I've learnt. I understand now why Jain considers Gupta 'the most important man in the system, not me'.

So what *is* the reason for the tigers of Panna constantly straying out? Is this natural behaviour? Tigers, like all carnivores, need to move; they need to disperse. That is why preserving wildlife corridors that link one national park to another are important. A wildlife corridor has been defined as a 'linear landscape element, which serves as a linkage between historically connected habitat/ natural areas, and is meant to facilitate movement between these natural areas'.

Corridors facilitate animal dispersal from isolated habitats and help counter biological processes that lead to species extinction. The critical features of a wildlife corridor are not its physical traits, such as its length or width or vegetation, but rather, how well a particular piece of land fulfils several functions, like the survival of species, mate finding, genetic interchange, propagation of plants, facilitation of travel, migration, movement of populations in response to environmental changes, natural disasters, and re-colonization of habitat areas by individuals. Animal movements are important for fitness, reproductive success, genetic diversity, and gene exchange among populations. There is, therefore, nothing unusual about the movement of Panna's tigers. What *is* unusual is the haphazard darting of the animals and bringing them back to the confines of the park. While Gupta may be doing a fine job, is this the right management decision to take? I pose this question to Raghu.

'Of course, dispersal is important. That is why we don't have fences for our protected areas. We want dispersal because we know connectivity is important but there is a lack of proper understanding of natural history; in fact, of the very basics of the natural history. I am amazed that we have managed what we have despite all this incompetence that exists in the system. They [the forest department] don't want conflict. How that can be made possible by catching and bringing the tiger back, I have no idea. Nor do I know why they think that animal will stay back.'

It seems like a cat-and-mouse game, the tigers moving out of Panna and the park's Dr Dolittle heading out to tranquillize and bring them back. From the department's point of view, the motivation may be to manage any potential conflict or harm to human lives when the animal 'strays'. That may explain the pressure to bring back the animal each time.

Meetings done, I am now keen to enter the forest. And next morning at 6 a.m., after a bitter cup of coffee, I set off in an open jeep, accompanied by a naturalist and our driver. We pass the ruins of a palace that was once the home of the maharaja of Panna, go past the Ken, and then embark on a safari into tiger territory.

Some two hours later, rejuvenated from my safari, I make one last stop for the day. I am on my way to a school that's run specifically for children from the Pardhi community that lives in hamlets around the park. The Pardhis are a traditional hunting community who were held responsible for the poaching of the tigers inside Panna.

Predominantly settled today in MP, Maharashtra, and Karnataka, Pardhis enjoyed pride of place in hunting parties for Mughals, zamindars, and others during the British Raj because of their unparalleled tracking skills. They moved from forest to forest, hunting smaller animals, wild boars and rabbits, but as the wildlife trade cast its net far and wide in India, the pressures (and the easy money therein) of hunting big cats became their primary occupation. In 1871, the British rule passed the Criminal Tribes Act, naming the Pardhis, Bawarias, and many such non-pastoral hunting communities as criminals. Even today, there is a lot of stigma against hunting communities and they are socially ostracized wherever they go, making employment or education impossible for them. The fact that this tribe was held responsible for depletion of tiger numbers didn't help its reputation either. International wildlife-trafficking rings relied on the Pardhis' skill at hunting the big cats and delivering tiger skin, meat, and bones for sale outside India, gradually emptying the reserves of all its tigers. A 2012 study done by the Tata Institute of Social Sciences in Mumbai found that 61 per cent of Pardhi children were never enrolled in school, making it hard even for the next generation to switch to other sources of livelihood.

In 1995, the World Conservation Union (IUCN) approved internationally accepted guidelines for the reintroduction of species. The guidelines clearly mention that 'there should be a good understanding of the reasons for the original decline and disappearance of the species considered for translocation and the causes of their reduction or elimination from the site(s) proposed for establishment of the species'. In other words, even before a reintroduction is planned, it is vital to eliminate the factors that drove the species to extinction. In this case, until the issue of rampant poaching was addressed, all efforts to rewild Panna with tigers would be futile. If the poaching had to be tackled, working with the Pardhis was a must. Park

Director G. Krishnamurthy mooted the idea of a hostel for Pardhi children in 2007.

Subsequently, Srinivas Murthy, an intrepid forest officer who was the field director of Panna from 2009 to 2015, consolidated these ideas further. Murthy was instrumental in managing the tiger translocation, along with introducing a host of relevant interventions to make it successful. He was eventually transferred out of the park because of a bold decision he took to oppose the government's ambitious river-linking project that would drown Panna—but more on that later. His decision to help set up a school for Pardhi children indicates a shift in thought on part of a department from policing to wanting to work with communities. Whether the department is equipped to handle complex community processes is another question, but the effort to set up the school suggested that at least it was becoming sensitive to such issues.

I walk into the school with a volunteer from the Last Wilderness Foundation, an NGO that performs the vital task of working with children living around protected areas. The NGO has been extending help to the Pardhi school by running vocational courses for the children here. I meet Roshni D'Souza, a young anthropologist from Mumbai who has been living in Panna for the last eight months and helps out at the Pardhi school. Secretly,

I am very impressed that this young woman has given up her fast-paced life in the city to work here in rural Madhya Pradesh.

I am taken to the girls' school. Roshni informs me there is also a school for the boys, but that's located in another village. We enter the compound, just a block of cement surrounded by a wall and an overgrown yard. The residential school houses around 30 girls from the Pardhi community. The youngest is around 6 years old, the eldest around 15. There is a small kitchen. A big room serves as their dormitory; every girl has her own trunk, a bed, and a mosquito net. The rooms are quite bare, with an occasional drawing or painting by one of the kids hanging on the walls. Otherwise, the facilities are quite basic, and the entire compound has no trees, plants, or even a playground.

The children have all been lined up for my visit, a fact that makes me feel a bit uneasy. A young girl in a yellow salwar kameez, egged on by Roshni, gets up and chants a well-known Hindi nursery rhyme. Roughly translated, it goes, 'A fish is the queen of water; if you take her out she will die.'

Everyone cheers the girl, who smiles shyly at me. I ask Roshni what happens to these girls once they go back to their homes. She says, 'Right now, we are trying to enrol them in some training workshops, such as stitching and tailoring, so that they can earn some money.' But she also informs me that apart from educating these girls, the forest department has been unable to do much. The Pardhi school is a good idea, but will it meet any social goals? I don't know. Removing these girls from their families must be heartbreaking for so many of them, some whom are as young as 5. I am saddened by the school, but more by the opportunity lost, of involving the Pardhis in conservation goals. Could their animal-tracking skills, for instance, not have been used by the forest department? Instead of trying to 'reform' a community and 'mainstream' them as tailors/sweepers/electricians, could their inherent abilities not have been used *for* conservation? I mull on this, drawing on my own experience of having worked with a hunting community near Sariska Tiger Reserve in Rajasthan. There, a community known as the Bawarias had been hounded and socially ostracized, not just by the land-owning community but also the forest department. They were believed to have poached all the tigers into extinction. It is easy to say that hunting communities need to be reformed, but it is far tougher to actually put those words into action and provide alternate livelihoods to people who already live on the margins of society. That is the rethink that any intervention with the Pardhis so desperately needs.

So is setting up schools sufficient to ensure that the Pardhi children don't revert to hunting ? I wonder as I leave the school with a heavy heart, the words of the poem echoing in my heart: 'A fish taken out of water will die.'

The bright side, though, is that thanks to NGOs like the Last Wilderness, this attitude is gradually changing. Roshni D'Souza and Bhavna Menon, managers at the NGO, have been working closely with the Pardhis, helping them find jobs once they leave school. Menon also organizes walking trails led by members of the Pardhi community with tourists to showcase their age-old knowledge of the forest and animal-tracking abilities. The endeavour has not only helped the Pardhis earn an income but also achieved an image makeover for the community.

Tracking radio-collared tigers is just one part of the story. There's also the task of protecting the forest, surrounded as it is by a sea of humanity. At one time, Panna had as many as nine villages inside; most of these were relocated outside its boundaries, and till 2017 only three remained in the reserve. Badhaun village was moved out in 2012, under an ambitious scheme introduced by the National Tiger Conservation Authority. The entire village is now closer to the town of Panna, not far from the forest office. With mud-washed homes painted a cool blue, the settlement here could have been the prototype of a happy village—people who moved out to make way for the tiger. But relocating humans is fraught with problems; it's never so simple as just a physical relocation. Premilal, a middle-aged man from the Yadav community, says he misses forest life. 'We lived inside Panna. My father, my grandfather were born there. I don't know why we moved out. Yes, we have a home here, but nowhere to graze our cattle.'

Seated next to him, 40-year-old Bhagirath from the Gond tribal community states that life outside the forest is tougher. He shares the cold mathematics of relocation. 'We got around Rs 835,000 as compensation. This house you see cost us Rs 30,000–50,000. Then we had to buy land for agriculture, which is quite far from here. We purchased 2 bighas of land for Rs 50,000. So you can see how much we are left with. Yes, money was given, but there are daily challenges.' His wife Imarti complains about having to walk longer hours to collect firewood. She admits she sneaks into the reserve, living under the constant fear of being caught by the forest officials. 'If I get caught, they will seize the firewood or any fodder I may have collected and just burn it.'

In spite of these problems, Bhagirath admits that accessibility is easier outside the forest, especially when there is a medical emergency or for sending his young children to school. Ironically, men like Bhagirath and their families were moved to make way for the tiger; today, a proposed multi-crore irrigation project will submerge the same land. Worse still, an Environmental Impact Assessment (EIA) report for the Ken–Betwa inter-link project states that more than 5,000 people will move into labour camps to construct a dam to link the rivers that give the project its name. The translocation of a seed population of tigers and the establishment of a healthy breeding population just seven years after the poaching debacle could have been Indian wildlife conservation's best success story—but for this new emerging threat.

In 2015, a year after the BJP came to power at the centre, it wanted to revive a proposal made by Atal Behari Vajpayee in 2004, to link the Ken and Betwa rivers in the hope of providing water to the parched parts of UP that straddle MP. The project entails the construction of the Daundhar Dam deep in the heart of Panna Tiger Reserve, considered to be its core area. The requisite clearances are proceeding at breakneck speed, even as every committee set up by the government has advised against it, thus paving the way for drowning 4,000 hectares of the forest.

It is perhaps also why Srinivas Murthy was booted out, for having red-flagged the move to link the Ken and Betwa rivers, highlighting that it would be detrimental for the park. He was subsequently transferred to MP's Kuno Wildlife Sanctuary.

Pegged at more than Rs 10,000 crore, the river-linking project involves the construction of a 77-metre-tall and 2,031-metre-wide dam at Daudhan (National Water Development Agency website, Government of India). A 221-km-long canal will also be built to transfer the water from the Ken basin to the Betwa basin for irrigating an estimated 6.35 lakh hectares of land in parched Bundelkhand, which straddles UP and MP. The intentions may be noble, as stated in every government document, but river linking is playing with nature—and the issue is more complex than being for or against the project. Even the EIA report outlines huge social and environmental costs, but still goes on to praise the project by stating 'the high adverse impacts are to a great extent offset by the positive impacts of large area getting the benefit of the command'.

There are others who disagree. Himanshu Thakkar of the South Asia Network of Dams, Rivers and People (SANDRP) has conducted an exhaustive critique of the EIA report, including the process by which information was shared with the public. In 2014, when the public hearing was conducted, Thakkar noted on the organisation's website: 'The public hearings required for the Ken Betwa River linking project Plan should be put up on the website of the Pollution Control Board a month before the actual public hearing. The full EIA and EMP [Environment Management Plan] were never put up on the MPPCB [Madhya Pradesh Pollution Control Board] website.' Thakkar further observed: 'The Ken Betwa Link project is a joint project between Uttar Pradesh (UP) and Madhya Pradesh (MP), about half of the benefits and downstream impacts in Ken and Betwa basins are to be faced by Uttar Pradesh, but the public hearings are not being conducted in UP at all, the proposed public hearing is only in MP! Even within MP, the link canals will pass through and thus affect people in Tikamgarh district, but the public hearing is not being held in Tikamgarh district either.'

Thakkar points out crucial mistakes in the EIA report. For instance, the report states in its executive summary that '[n]one of the species of aquatic plants come either under rare or endangered or endemic or threatened categories (REET)'. In fact, scientific studies clearly show that the Ken River has at least four endangered and nine vulnerable species. The EIA also keeps mum about the Ken Gharial Sanctuary in the downstream area, which will be completely destroyed because of the project. There are a number of glaring errors in report that show, according to Thakkar, the 'ecological illiteracy of the project proponents'.

Environmental lawyer Ritwick Dutta, who is challenging the entire process in the Supreme Court, has argued that the permissions given by the Standing Committee of the National Board of Wildlife are outside of its jurisdiction. The committee can only give permissions for projects that benefit wildlife. How, then, can drowning the tiger's habitat be of any benefit?

Besides Thakkar, a number of experts have written letters to the expert appraisal committee (EAC) of the ministry of Environment, Forests and Climate Change, asking them to revisit the terms of reference on the basis of which the project was approved. Brij Gopal, an aquatic ecologist, wrote a letter to the EAC on 17 January 2016 stating, 'The proposal for K–B linking is based on the assumption that River Ken basin is water surplus that is not supported by any hydrological data, is erroneous, and farthest possible from the reality on the ground. The proposal is rooted in the desire to construct four projects in upper Betwa basin for growing rice and sugarcane in place of pulses and millets, while Panna district gets no water from the project.'

Could there have been any other alternatives that could have been explored, given the huge financial, ecological, and human costs of the project? Focusing on less water-intensive crops and/or reviving traditional water bodies like tanks and village ponds that were once part of the region could have been ways to bring water to a dry region that would not even need taking a loan, chopping 7 lakh trees, or flooding two sanctuaries. In spite of the caution raised by several water experts and wildlife biologists, it is clear that the governments at the centre and state are in no mood to listen. In May 2016, the standing committee of the National Board for Wildlife gave a go-ahead to the project despite the fact that the site-inspection report was still not complete. In March 2017, the forest advisory committee gave a conditional green light, if several concerns could be addressed, such as reducing the height of the dam to save some of the wildlife habitats from being flooded. As of 2019, the Wildlife Clearance issue was being discussed in the Supreme Court. The Project does not have clearance under the Forest (Conservation) Act, 1980. Therefore, no activity can commence on the ground till these clearances had been sought.

As the tussle between environmentalists and proponents of the ambitious scheme continues, the biggest loss will be the tigers'. Once the Ken–Betwa river-linking project comes through, the tiger reserve will be bifurcated by a dam, leaving even less forest land behind. It seems ironic that on the one hand, the state government has worked so hard to bring tigers back to Panna, and on the other, it is now working overtime to ensure that the tiger habitat is submerged by a dam. The efforts to bring back Stripey in Panna may have been successful, but if the political will is inclined towards a dam at the cost of the tiger then even the best rewilding projects can get derailed.

3

This Little Piggy Found His Way Back Home

◎

COMMON NAME: Pygmy hog [English]; Takuri bora [Assamese]
SCIENTIFIC NAME: *Porcula salvania*
CURRENT CONSERVATION STATUS: Critically endangered
(200–500 individuals)

◎

On the edge of Guwahati, surrounded by a grove of sal trees, the legacy of British naturalist and author Gerald Durrell lives on. Open the gates bearing the logo of the dodo, and you will find a species being saved from annihilation. Backed by some of the best scientific minds in the world, wildlife biologist Goutam Narayan and veterinarian Parag Deka are making sure the smallest pig in the world doesn't slip down the extinction vortex.

The Pygmy Hog Research and Breeding Centre is the only site in the world where these animals are being bred in captivity with the objective of rewilding them. A peek inside the enclosures—typical 'Durrell' ones that are replicas of the animal's natural habitat—and you notice that there's tall grass everywhere. You have to patiently look past the abundant *Narenga porphyrocoma* and *Imperata cylindrica* collected from original hog habitats in Manas National Park and Bornadi Wildlife Sanctuary to spot the animal here. If it's difficult to do so in captivity, imagine how tough it must be in the wild. But when the sun's rays shine their full glory on the enclosure, that streak of black amid the green is a giveaway. A sow comes out, followed by two piglets. All three have their noses to the ground, as if that is their assigned homework for the day—at least until the piglets begin chasing each other around the grass, returning to the mother for a quick drink of milk.

I spend over an hour watching the pigs' antics before trudging to the top of the hill to meet one of the men who's made this possible—Goutam Narayan. Funnily enough, Goutam had studied to be a birdman. He was working on the highly endangered Bengal florican in

Manas when the pygmy hog caught his attention. (About 150 kilometres from Guwahati, Manas National Park was the last stronghold of the pygmy hog.)

Goutam tells me he's a great admirer of Gerald Durrell, the British naturalist whose hilarious *My Family and Other Animals* caught the attention of the world. Durrell went on to inspire an entire generation of biologists when he set up the Jersey Wildlife Conservation Trust in Jersey, the largest of the Channel Islands between the UK and France. So when Goutam got an opportunity to work for Durrell's organization to set up the Pygmy Hog Research and Breeding Centre as part of the Durrell Wildlife Conservation Trust, he seized it. He moved from the Bombay Natural History Society to set up the breeding centre in Assam. Today, Goutam is considered to have done more for the gentle pig than any other person in the world.

Incidentally, Durrell was born in Jamshedpur, and it is only right that his legacy now lives on in the country of his birth, in the form of India's first captive breeding centre. The pioneering project is perhaps the only one of its kind to have bred an animal species in captivity on this scale and released it into the wild. It is also one of the first projects in India that recognized captive breeding as a conservation tool. That's why the project is among the top 10 rewilding projects from across the world that have been chosen for Durrell's centenary celebrations, in 2025. It's a major milestone for the programme—and for the species. But why did this icon of India's grasslands move to the critically endangered list in the first place?

The pygmy hog is by far the smallest and rarest wild pig in the world. Once quite widely distributed through the Terai–Duar tract of alluvial grasslands along the Himalayan foothills, the species is now confined to Manas in northwest Assam, along the state's border with Bhutan, and in protected areas like the Sonai Rupa Wildlife Sanctuary and the Rajiv Gandhi Orang National Park. Assam could well be the last stand of this shy animal. The IUCN Red List has accorded this species the highest conservation priority rating of all members of the pig family, and it has long been considered to be among the most endangered of all mammals. So when did the trouble first begin for these pigs?

The species was first recorded in 1847, though very little was known about it: pygmy hogs were never reported to be plentiful to begin with, and by the 1960s, the species was already feared extinct by naturalists such as E.P. Gee. Around this time, Jersey resident Captain Tessier-Yandell contacted Gerald Durrell, informing him that he was going to Assam to manage a tea plantation. Durrell requested the Captain to see if small populations of the hog existed in the region. In the early

1970s, much more information became available, after the species was rediscovered in some reserve forests across Assam, such as in Bornadi and Khalingduar.

In 1971, extensive fires drove pygmy hogs out from their grassland homes and into a tea garden in the area, and a tea manager of one of the estates captured and maintained several of them. When he learnt of this, Jeremy Mallinson, Durrell's right-hand man who had helped to set up the zoo, visited Assam to proffer advice on their husbandry, and several litters were born. William Oliver, a research assistant of the Jersey Wildlife Conservation Trust, carried out extensive fieldwork during 1977, the results of which were published in the Trust's *Special Scientific Report No. 1*. Between 1971 and 1977, at least 45 pygmy hogs were acquired and at least 40 were born in captivity. But by the mid-1980s, none had survived. There was clearly a lack of experience in managing the species in captivity. By then, it had also become clear that the species had almost certainly disappeared from the wild and urgent action was required to save it.

In 1985, following further Trust-sponsored visits, William Oliver was invited by the government of India to submit an action plan for the species' conservation. This was formally accepted in 1988. The plan included the initiation of a properly structured co-operative

51

breeding programme and more long-term field studies of the behaviour and ecology of the species. The wildlife division of Assam's forest department was charged with implementing the recommendations, but failed to do so. In November 1991, a revised plan was submitted, and three years later, a recovery programme was finally approved under the auspices of a new International Conservation Management and Research Agreement for the Pygmy Hog. This agreement, the first of its kind in India, was signed in February 1995 by the ministry of Environment and Forests, the Indian government, the State Forest Department of Assam and the Durrell Wildlife Conservation Trust (then Jersey Wildlife Preservation Trust). Slow as the government was in its response, things finally came together for the pygmy hog, and the captive breeding centre was set up in Basistha near Guwahati in 1996. What is noteworthy is that the governments at the centre and in the state recognized the value of a captive-breeding project and were willing to commit the resources for it in terms of funds, manpower, and land. In doing so, they displayed tremendous foresight and vision, which helped the breeding programme immensely. But enabling policies was one thing; a breeding programme that would help populate the species in the wild would have its own set of challenges.

There was a larger question here that needed to be addressed: why was the pygmy hog so vulnerable to changes in its external environment?

Pygmy hogs are dependent on, and specifically adapted to, undisturbed patches of grasslands, typically comprising dense tall grass swards mixed with a wide variety of herbaceous plants and early colonizing shrubs and young trees. These are crucial for the survival of a number of other grassland species, such as the great Indian one-horned rhinoceros (*Rhinoceros unicornis*), swamp deer (*Rucervus duvaucelii*), hog deer (*Axis porcinus*), water buffalo (*Bubalus arnee*), hispid hare (*Caprolagus hispidus*), and the Bengal florican (*Houbaropsis bengalensis*). However, given that it has disappeared from many protected areas that still support remnant populations of these other species, it was clear that the diminutive hog has a much lower tolerance to human-induced changes to its habitat and other disturbances. Historically, these grasslands were probably maintained by changing river courses as well as by grazing and trampling pressures from large herbivores such as rhinos, elephants, buffalo, and deer. With the extermination or sharp decline in large wild herbivore populations, this control mechanism had more or less vanished, thus endangering the hog. And then there was the biggest threat—fire.

The age-old custom of indiscriminate dry-season burning through 'fire lines' that the forest staff practised to manage or control the spread of potential forest fires, coupled with the department's tendency to plant trees everywhere to rewild a habitat, wreaked havoc for grassland species like the pygmy hog. (A fire line, also called fire break, is the practice of burning a strip of vegetation and clearing the land so that the flames don't spread in case of a fire.) Large-scale annual (in some areas, biannual) burning in protected areas, also practised by local communities for the commercial harvesting of thatch, destroyed the surface litter in which the hogs forage, along with the ground fauna (for example, insects and annelids) that were an important source of food. This also exposed the surface substrate, which typically becomes hard prior to the rains, making the establishment of roots more difficult and less profitable. This unscientific management of grasslands was the primary cause for the decline of this sensitive species. Until this threat was addressed, no captive-breeding programme could be a success. That's why it was imperative that the pygmy hog conservation team work closely with the sanctuary authorities, especially in the areas identified for releasing the captive-bred animals so as to improve the protection and management of these areas and control the annual dry-season burning of grass.

The sanctuary staff was also trained in wildlife monitoring and habitat management to help in the restoration of the grassland habitat and keep an eye on the hogs post their release. That is perhaps the biggest strength of the pygmy hog project: unlike other programmes that get stuck in the first phase—the breeding of the animals—this worked on the causes of decline of the species as well as on addressing those threats.

Once the permissions had been obtained from the government, Goutam and his team had the onerous task of capturing the animals from the wild to serve as a founder population for the breeding programme. Since this was a critically endangered species, they had to ensure there were no mortalities in the capture phase. In 1996, six wild hogs were caught from Manas and transferred to the centre at Basistha. The method of capture was rather unconventional: the team walked through the grasslands with captive elephants and forest personnel and collected the pygmy hogs scattered amid the grass in bags. It was a tense time and a monumental task, recalls Goutam.

At the time of their capture, three adult females were pregnant, and produced healthy litters, so the captive population increased to 18. Seven more litters were born the following year and the captive hog population, at 35, almost doubled in 1997. Similar success in captive breeding in subsequent years saw the population at Basistha rise to 77 in 2001, a thirteen-fold increase in the stock in four years. This unanticipated and rapid increase in the captive population necessitated rigorous curbs on their reproduction. A population of around 70 hogs was maintained in captivity until 2007, and the first releases were conducted in 2008. Veterinary doctor Parag Deka, who had trained at Durrell's Jersey Zoo, closely monitored the hogs' diet and health indicators before they were released in the wild.

Releasing the animal in the wild is never as simple as opening the enclosure gates. It requires meticulous preparation—of the animals as well as the habitat they are to be released in. The captive-bred animals were trained for life in the wild right from their birth.

Social groups of unrelated and mostly young hogs were integrated at Basistha before being transferred to a specially constructed pre-release facility in Potasali, on the outskirts of Nameri National Park, east of the Sonai Rupai Wildlife Sanctuary. Every effort was made to 'precondition' the animals for their eventual release by maintaining them in three separate social groups. They were kept in simulated natural habitats intended to encourage natural foraging, nest building and other behaviours, while minimizing human contact to mitigate tameness and other characteristics consequent upon their captive management. Radio harnesses designed for post-release monitoring studies were also field-tested by trial attachments to two hogs in each group, but unexpected problems arose in their long-term use. If they were too tight, they were prone to cause serious skin lesions, and if too loose, the hogs were able to escape from the harnesses. Such setbacks made it difficult for Parag and his team to monitor the pigs released in the wild. In the interim, alternative means of monitoring had to be put in place, like camera trapping or screening for field signs, such as forage marks, tracks, faeces, and nests.

After a five-month stay in Potasali, these hogs were transferred in early May to temporary 'soft-release' enclosures constructed for this purpose in a relatively secluded but easily accessible natural habitat in the far interior of Sonai Rupai. These enclosures were also rigged with two lines of electric fencing and kept under continual surveillance as a precaution against potential predators and to deter incursion by wild elephants. The hogs were maintained for three days in these enclosures before being released, by the simple expedient of removing sections of fencing and allowing the animals to find their way out.

Sixteen pygmy hogs (seven males and nine females) were released in the Gelgeli grasslands of Sonai Rupai in

May 2008. Indirect evidence such as droppings and nests suggested that at least 10 to 12 continued to survive several months after release. Footprints of newborn hogs, too, were seen, indicating successful farrowing in the wild by a released female. A video-camera trap was deployed near active nests. The hogs caught on camera appeared healthy and had shiny coats, unlike the somewhat emaciated hogs captured from the wild in Manas in 1996. Some of them were identified by hair-clipping marks, as they had been shaved before release. That the hogs appeared to be in good health nine months after their release, despite harsh weather and sometimes difficult foraging conditions, was most encouraging for the project team. It not only confirmed their survival, but suggested their successful adaptation to the wild after at least one or (in most cases) two generations of captive management. Following similar protocol, 9 hogs were released in May 2009 and 10 in May 2010, thereby releasing a total of 35 hogs in different locations of the Gelgeli grasslands in Sonai Rupai.

Wildlife filmmaker Kalyan Varma, who has been associated with the project for the last several years, was fortunate enough to be present when the 100th hog was released into the wild. 'I was sitting with Parag at their breeding centre at Potasali and I suddenly heard a yelp from him. It was August 2017. He came out of his room screaming, asking me to come have a look at something he had just discovered. He held my hand and led me to his laptop. He showed me a slideshow of many photos of camera traps and paused at one. "Do you see it?" he asked. I said. "Yes! That's a pygmy hog with two young ones. Is this the mother you released?" He jumped and said, "No and that's why I am so excited." The photo he showed me was of a pygmy hog without any ear tag. All the hogs that are released into the wild by the centre are nipped in the ear and this one was not. So it meant that we were looking at the second and third generations of pygmy hogs born from the released ones. I have never seen anyone so happy. "I can die now," he said. "I can say I have done something to save a species in my life."'

Varma further noted in an email to me:

'I think that spirit is what really drives the pygmy hog conservation programme. I have seen big breeding/relocation projects with lots of funds and staff, but what we are looking at is one of the most [successful] conservation stories, which was executed by two guys and a half a dozen field staff. Surely there is nothing like it.'

Even as the programme has grown over the years, the species is still not out of the red. The main difficulties in the implementation of this project are largely related to a serious shortage of a suitable habitat for reintroduction. Although most of the remaining pygmy hog habitats are in protected areas, they have suffered from unscientific management and flash floods caused by natural or artificial dams among other setbacks. While it was heartening to witness the remarkable fruits of a meticulously planned conservation programme, I wanted to see the pygmy hog in its natural environment—and learn why the grasslands were in such danger.

I choose for my site visit the Rajiv Gandhi Orang National Park, which is largely swampy with thick grasslands and is situated on the north bank of the Brahmaputra. The captive-breeding programme has been particularly successful in Orang, where the population has increased to over 100 individuals following the release of around 59 hogs over a period of 5 years. Narayan and his team have worked hard here, and I've also been told that some of the best grassland habitats still persist in the park, which could perhaps be a reason for its success. I haven't chosen the best time of the year to visit; it's the day of Bihu, considered to be one of Assam's biggest festivals, a time when the state virtually comes to a halt. A big feast is held to thank the gods for a bountiful harvest after the crop has been cut and the granaries are brimming with food grains, warranting a reason to celebrate. The celebrations are usually marked with a massive bonfire followed by a grand feast. As I make my way to the park on a bleak winter evening, there's a flurry of activity everywhere as I pass through small villages and towns. Ducks, poultry, and fish are in high demand: the men balance the live dinner on bamboo cages and cycle home to begin preparing for the feast.

I stay at a dilapidated tourist lodge where the plan is to halt for the night. Sadly, not too many tourists make their way to Orang, so the government-run lodge with its paan-stained walls and torn curtains is the only facility available. There is a forest rest house inside the park, but that is reserved for senior officers, I am told by the local DFO (divisional forest officer). While all of Assam may be feasting, at the lodge there isn't any dinner—only leftovers from the afternoon—as the cook has gone home to celebrate. I fall asleep with the hope that the forest will cheer me up, with perhaps a sighting of a pygmy hog if I'm really lucky.

I reach the gates of the park just before sunrise; inside, the thick green foliage is covered in white mist, making visibility poor. Since it's the day after the festival, everyone has slept late and I am unable to find a forest guard to accompany us inside. While I am primarily here to witness the genteel pygmy hogs, I keep a lookout for the rhino as well. This is, after all, rhino territory. The pachyderms, though, are known to be temperamental, so I am told I must wait for the forest staff to take us inside.

Thirty minutes later, accompanied by a forest guard, we make our way inside. What the park lacks in tourist facilities, it more than makes up for with its natural splendour. The sun is just about rising, opening up dramatic vistas of a rich tapestry of swampy grasslands and *beels* (wetlands). The jeep tries to push past the giant grass; this is Jurassic Park revisited. It's as though nature's tentacles have resisted the march of time. I understand now the physical challenges Goutam and his team must have endured looking for pygmy hogs in this landscape. I expect a prehistoric dinosaur to come rushing out any moment. The driver is surprised we want to see the pygmy hog—'people come here to see the rhinos'. I ask him if he has seen pygmy hogs in all his years here. 'Sometimes.' He replies, looking through his binoculars for rhinos. 'They are not that easy to spot.'

The quaint little guesthouse is a welcome stopover, and provides a panoramic view of the entire park once the fog lifts. From this vantage point, we can see a rhino munching on his morning breakfast down below. A herd of barking deer arrives, bowing their slender necks every few minutes and then looking up suspiciously, ever alert for a predator.

I sit in the sun chatting with Ghanshyam Rajvanshi. The forest guard has written a book in Assamese about his life in the jungles and regales me with his adventures. He is a born storyteller, and tells me about his encounters with wild tigers and rhinos during his stint at Kaziranga—testimonials to the threats the front-line staff face each day—with great humour and suspense. I feel small, waiting for that one fleeting glimpse of a pygmy hog or even a tiger's tail; this man personifies a lifetime spent in the wilderness. He is the one who deserves to tell a story about the jungle, not me.

We drive around for another hour, making our way through the near-impenetrable tall grass. Pygmy hogs are extremely shy, which is why very few observations of their behaviour in the wild exist. The visit to Orang makes me realize that the story of the pygmy hog is as much about the need to preserve the grasslands as it is about the animal. Outside the protected area, grasslands are being destroyed at an alarmingly high rate, as communities continue to use it for thatch. Working with the locals, therefore, could be a way forward for creating newer habitats for the hog.

Until they released the hogs, Goutam and Parag—and the rest of the world—knew little about the behaviour of pygmy hogs in the wild. Their initial attempts to study the animals met a few hurdles, but as Parag points out they tried to work their way around them in each instance. For instance, implants were tried on seven hogs in 2011 and 2012, but had to be discontinued because of the associated health problems and the limited battery life. With further advancements in technology, it may be possible to experiment with devices that can be inserted into the hog's body without any health problems. Meanwhile, monitoring the animals at the captive stage and even after their release threw up a wealth of information. I learnt from Goutam that pygmy hogs are the only pigs—and one of the few mammals— that build and use nests all times of the year, and not just for farrowing. Their nests, in turn, were a mine of information for researchers. The number of nests and their distance from each other, among other things, shed light on the animals' social life. Through radio telemetry, it was observed that each released social group generally stayed within 150 metres of an active nest. The group remained around the nest for around four to six weeks before moving to another site and building another nest. Sometimes, nests were abandoned because the site was disturbed or from the threat of wild elephants.

Pygmy hogs are largely diurnal and rarely emerge in the dark. Captive animals were seen emerging in the dark only when they found their favourite food, termites. Studies have shown that the released hogs would disperse up to 3 kilometres from their release site, which was a good sign as it meant they were not living in a psychological cage and were reclaiming their habitat in the true sense. The team learnt that hogs generally avoided flooded or very boggy areas, or those that saw intense activity from wild elephants or wild pigs. There was also enough evidence to suggest that the released hogs have given birth in the wild, indicating that the programme has been a success.

In the years it's been active, the Pygmy Hog Conservation and Breeding Programme has helped release 21 social groups comprising 100 individuals into three protected areas of Assam. There are now plans to take the project outside of Assam, to West Bengal. For Parag and Goutam, the learning curve has been sharp, and they have realized that it takes years, if not decades, of persistent efforts to implement a successful recovery programme. The efforts to breed and rewild the pygmy hog will go from strength to strength, because the aim of the programme has always been to ensure the animals make it back to the wild. From the design of the enclosures through to the release protocols, the emphasis has been on how the pygmy hog will make it back home. The only lacuna I found was the limited interaction with the local community, but that's common to many conservation projects in India. Involving the communities, if not directly in saving the pygmy hog, then at least in grassland protection may be a necessary long-term measure to ensure that this little piggy need no longer be dependent on a captive-breeding project for its survival in the wild. That should be the ultimate goal of any rewilding project. And luckily for the pygmy hog, it is one step closer to that goal.

4

How to Raise Baby Turtles and Gharials

◎

COMMON NAME: Red-crowned roofed turtle [English];
saal [Hindi]

SCIENTIFIC NAME: *Batagur kachuga*

CURRENT CONSERVATION STATUS: Critically endangered

◎

'It's time,' announces Santram, immersing his hand into steaming hot sand baked in the 45°C heat, groping for the baby. *Babies*, actually, for there's more than one in this case—several will emerge from the earth's womb they have been incubating in, in this makeshift turtle nursery by the side of the Chambal River.

Santram gently lifts a baby in the air. It's not possible to tell at this stage if it's a boy or a girl. But for a species that has not more than 500 adults left in the wild, every individual matters. The creature that has just made its way out of the sun-baked sand is the critically endangered red-crowned roof turtle. The male has striking red stripes on its head and neck, assumed to attract females. This isn't the only occupant of this turtle nursery, though: the three-striped roof turtle (*Batagur dhongoka*), also struggling for survival, is a co-habitant.

Until a few years ago, Santram used to earn a living by selling turtles and turtle eggs—many of which are shipped out from India's porous borders to markets in Southeast Asia—to poachers by the kilo. He now plays foster parent to more than 200 nests along the banks of the Chambal in the National Chambal Gharial Wildlife Sanctuary that straddles UP, Rajasthan, and MP. (The sanctuary, though home to nine species of turtles, is, as the name suggests, actually famed for its crocodilian species—the Indian mugger and the Indian gharial.) He guards these nests against predators like stray dogs or jackals, and sometimes even dacoits. 'Dacoits are harmless, really. They usually just ask us to cook them a meal and go away quietly.'

As he talks, Santram eases the hatchlings out of the sand and pops them in a plastic bucket. The newborns, clambering over each other, sound like corn popping. The hatchlings are impatient; they want to rush to the cool waters of the Chambal, their home. But Shailendra Singh, scientist with the Turtle Survival Alliance, takes them on a slight detour. 'Not so fast,' he tells them, as he readies a sharp needle. He must first delicately insert an injectable tag into the turtles. This tag assigns a number to each individual and records vital data—its length, weight, etc. This data will help scientists determine the life history of these species, about which so little is known.

A 'chelonian' (relating to tortoises and turtles) conservationist, Shailendra has been leading the effort to rewild turtles in India. As a child, he had two pet turtles, and would walk miles to rescue these creatures from village ponds if found to be in distress. As he grew up, this childhood passion found a constructive channel, and he began conducting extensive research and studying the population structure of a turtle species on the Gomti River (a tributary of the Ganga that flows through UP). He's now focusing his attention on ensuring that turtles are not wiped out from the Chambal.

About 350 species of tortoises and freshwater turtles are known to exist, representatives of a lineage as old as the dinosaurs. Around 54 per cent of these are likely to become extinct in the next few decades, with pollution, destroyed habitats, indiscriminate hunting, use in traditional Chinese medicines, and illegal pet trade among the main culprits. While not as obviously visible as birds or ecologically dominant as large mammals or fish, turtles often fulfil important roles in the ecosystem, including in vegetation management, control of insect populations, and keeping the water clean and healthy by scavenging dead animals. India harbours some 24 species of freshwater turtles and 4 species of sea turtles; more than half of these species are globally threatened.

In India, it is the Chambal landscape where efforts to save the red-crowned roofed turtles are on. This turtle, commonly known as the Batagur (*Batagur kachuga*, previously *Kachuga kachuga*), features in the list of the world's 25 most threatened turtle species compiled by the Turtle Conservation Fund (TCF) in 2003. Historically, this turtle was widespread in the main riverine sections of the Ganga in India, Bangladesh, and Nepal. It's also said to occur in the Brahmaputra basin, though this has not yet been clearly established by existing museum records. Its populations have collapsed in recent decades as a result of the combined impacts of degradation and destruction of critical nesting and feeding habitats, egg collection, consumption, entanglement in fishing nets, and pollution. Only a single substantial population is known to remain in the country, in the National Chambal Gharial Wildlife Sanctuary, along some 220 kilometres of the Chambal River, across Madhya Pradesh, UP, and Rajasthan. Even this protected population and its habitat are under threat from sand mining, hydroelectric infrastructure plans, collection, and accidental mortality.

At the nesting site, Shailendra has tagged and weighed the turtles. The bucketful of ninjas is now ready for release. A flick of the wrist tips the bucket at the water's edge, easing the hatchlings onto the sand. There's a mad scramble as the hatchlings race each other to the cool waters. Once they take to the water, it's a smooth glide to freedom. It isn't without its dangers, however: they may face numerous predators, on their journey. Turtles have evolved a life strategy characterized by slow growth and late maturity (usually on the order of 10–15 years), longevity, and a relatively modest annual reproductive output (1 to over 100 eggs per mature female per year, depending on species). But very low survival rates of eggs and juveniles makes the population vulnerable to poaching and habitat changes.

In 2017, Shailendra and his teammates were able to protect over 600 nests, and so helped hatch 9,622 baby turtles. Their overall hatching rate was an impressive 83 per cent. So *why* do the turtles need assistance through riverside hatcheries? Shailendra explains that the segment of the Chambal where the species resides is heavily impacted by an irregular flow of water, sand collection, riverside agriculture, and clandestine and banned activities such as fishing in a sanctuary area. Perhaps the biggest threat comes at the most crucial stage in the lifecycle of the turtle. Water from dams in Rajasthan is released in March–May, when the turtles are nesting or hatching. A huge gush of water inundates their nests, thus leading to mass mortalities. 'I have written to the state biodiversity board about this. In 2015, around 500 turtle nests were flooded. It got so bad that we had to quickly collect the eggs and shift them from the hatcheries. We are struggling to tell people that the red-crowned turtle and birds like the Indian skimmer on this stretch of the Chambal are more threatened than gharials,' he says.

Another concern, of course, are adult turtles— they are taken for their meat, for the illegal pet trade, or are victims of indirect trapping when thousands of line hooks are set on the river bottom to trap softshell turtles whose plastrons (undershells) are worth a lot of money on the black market for use in Chinese medicinal practices and as aphrodisiacs.

So little is known about turtles, and we are losing them before we can discover more, laments Shailendra. Having released the hatchlings, we make our way back from the sandy banks of the Chambal to the main highway leading to Etawah. Earlier in March, Shailendra was alerted by the police about the confiscation of 23 red-crowned roofed turtles, destined for foreign markets from two poachers linked with an illegal international wildlife trade network. The turtles were stuffed into suitcases, bound for the port in Chennai. Fortunately, the poachers were apprehended by officers of UP Police and the Wildlife Crime Control Bureau that is managed by the central government, and sent to a rescue centre in Etawah with help from the Turtle Survival Alliance and the state forest department.

At the rescue centre, the turtles are kept in shallow ponds. Shailendra gently lifts the turtles up to show us their injury marks and a fungal infection on their shell, developed when the hapless creatures were shipped to the rescue centre by the police. Their limbs were taped and they were wrapped in small boxes by the poachers when the truck was apprehended. He is hopeful that once the infection is cured and their quarantine period is over, they can be released back in the Chambal. In this setting, they can be carefully monitored for symptoms that may arise following the stress they have endured from their capture, packaging, and subsequent confiscation. With the species staring at near extinction, permission is being sought from the ministry of Environment, Forest and Climate Change at the centre to keep six males from this confiscated group for breeding efforts to provide critical new bloodlines to the existing populations. While the threat of poaching is still very much alive, the project team believes that it is in the best interests of the population of this critically endangered turtle to release the remaining individuals back into their native habitat. This would help maintain the genetic diversity and health of the remaining wild population. To help identify the individual turtles should they be seen again by researchers or, in a worst-case scenario, taken again by poachers, all the individuals will be inserted with a tiny microchip tag to identify them. These are similar to the ones Shailendra had inserted in the hatchlings before releasing in the Chambal.

Some scientists have raised the issue of whether the focus needs to shift from riverside hatcheries to better protection for adult turtles. Jeffrey Lang, professor emeritus at the University of North Dakota, USA, considers riverside hatcheries to be well intentioned, but insists that more serious science is needed to show how they are contributing to conservation. Lang points out: 'This procedure [of shifting the eggs to riverside hatcheries] could be [skewing] the sex ratios. In the same way, early work with sea turtles shifted the balance from female to male by transplanting eggs to new incubation sites. The adult turtles are the ones facing the real threats, and once the adult nesting females disappear, the population is doomed.'

And the threat of poaching is a serious one. A study conducted by Shailendra and his team—it appeared in the journal *Biological Conservation*—reveals the gravity of the situation for freshwater turtles. Using data from 223 enforcement seizure reports obtained through systematic online searches, the study shows that at least 15 of India's 28 tortoise and freshwater turtle species (TFTs)—including 10 threatened species on the IUCN list—are illegally removed from the wild. In fact, over 58,000 live individuals were seized during 2011–15. Nearly 90 per cent of all seizures were from illegal commercial trade and there were numerous reports of Indian TFTs being transported by road, rail, and air within India as well as to known pet- and meat-trading hubs in Bangladesh, Thailand, and other Southeast Asian countries. Within India, the Gangetic Plains accounted for 46 per cent of all seizures, with Lucknow and Kanpur being major hubs.

Fortunately, there are other heroes in the system that wildlife biologists have partnered with. For instance, additional superintendent of police (ASP) Aravind Chaturvedi from the Special Task Force on wildlife crime, Uttar Pradesh Police, was responsible for busting a number of poaching rackets. In January

2017, Chaturvedi made history when he brought down a poaching racket that involved more than 6,500 Indian flap-shelled turtles, considered to be the largest wildlife seizure that India has seen in recent times. Flap-shelled turtles are sold on the black market for the purported aphrodisiac quality of their meat, and these were being taken in jute bags to markets in Kolkata.

In an email interview, Chaturvedi writes: 'Since 2005, a dedicated team under my command started working on wildlife cases in a systematic way. Initially, I concentrated on tiger and leopard cases, but later realized that every animal is important in the ecological chain and contributes in its own way to it. During March–April 2018, my teams have recovered more that 200 kilograms of turtle calipees which was to be smuggled to Bangladesh via Maldah in West Bengal.' He confesses that wildlife crime investigation is his passion and nothing gives him more satisfaction than 'working for animals and nature'.

If it wasn't for the timely action of the Wildlife Crime Control Bureau, the UP Forest Department, and wildlife scientists like Shailendra, thousands of turtles every year would make it to the pet and meat trade, pushing them closer to extinction. Shailendra, Santram, and Chaturvedi are the unsung heroes of this story, for they have ensured the survival of turtles in the Chambal.

While turtles in the Chambal have seen some relief, the story is quite different for those in India's most sacred river, the Ganga. Once home to 13 species of turtles, rampant pollution has robbed the river of its chelonian species along most of its journey. In the 1980s, the government of India hit upon an ingenious plan under the Ganga Action Plan to not only help revive turtle populations but also help clean the river.

Not far from the Sarnath temple, where the Buddha is said to have gained enlightenment, a turtle-breeding centre was set up by the UP Forest Department. The plan was to breed species like Indian soft-shelled turtle and Indian flap-shelled turtle in captivity and release them in the wild once the hatchlings were big enough. The 'wild' in this case was a 7-kilometre stretch of the Ganga close to Varanasi that was declared a turtle sanctuary in 1989. No sand mining or boats were allowed in this part of the river in the hope of providing a safe haven for the turtles (as well as other species like the Gangetic dolphin that was once found here).

Under this plan, claims the UP Forest Department, during the period 1987–92, 55,690 turtle eggs were brought to the Sarnath Breeding Centre from the Chambal, out of which 30,091 newborns were reared. Of these, 28,920 were released in the Ganga. In 2005–6, eggs were again brought from the Chambal, from which 3,297 newborns were reared, and in March 2009, 1,549 of these were released in the Ganga. Based on these figures, one would assume the river would be teeming with turtles. But even after a decade-long breeding programme, parts of the river are still without turtles. No one knows what happened to the turtles post their release at Varanasi. None of the hatchlings were marked for identification and no annual census is taken. The breeding centre that could have helped rewild turtles lay mired in government apathy. While the breeding centre may not have done well, for the few remaining turtles that were still found in the Ganga, the sandy bar of the sanctuary served as an important nesting site.

However the Central Government had ambitious plans: the Inland Waterways Authority under the Shipping ministry declared its intention in 2016 to set up a waterway on the Ganga. In co-ordination with the ministries of Water Resources, Urban Development, Forest and Environment, Industry, Tourism, and Power, the plan was to dredge large portions of the river to develop a waterway along the Gangotri–Kanpur–Allahabad–Kolkata route for cargo and passenger movement. But the environmental consequences of this were not thought out well.

And then it got worse. In 2018, the state government submitted a proposal to denotify the turtle sanctuary. An entire wildlife sanctuary was about to be wiped off the map of India for the government's ambitious waterways project to come through. This proposal was accepted by the centre in order to facilitate the movement of ships along the stretch of the Ganga at Varanasi.

When the rumours were confirmed by an order from the state government, I thought the media would be screaming with headlines on it. But there was dead silence. I was one of the first journalists to report on the story, but initially no media house wanted to publish it. Finally, the online portal *The Wire* agreed to carry the story with the supporting documents that showed that the government had indeed denotified the sanctuary.

Two days after the story broke online, Prime Minister Narendra Modi received the UN Environment Champions of the Earth award, the organization's highest environmental honour, from Secretary General Antonio Guterres in New Delhi. The PM had been recognized for his 'exemplary action on climate change and sustainable development'.

The tiny turtles, meanwhile, had paid a heavy price. In a small way so did I. A job that I was interviewing for did not work out and the bosses went silent on me. I was later told (off the record) that I was seen as a 'trouble-maker' for having broken the story. The government, meanwhile, announced that they would compensate for the denotification of the sanctuary at Varanasi by forming another near Allahabad.

In the tussle between turtles and mega ships, the government has given preference to the latter. While Shailendra and his team have been successful to some extent in protecting turtles in the Chambal, the fate of the chelonians in the Ganga hangs by a perilous cord. Politics, it seems, won the day over the turtles.

◎

COMMON NAME: Indian gharial
SCIENTIFIC NAME: *Gavialis gangeticus*
CONSERVATION STATUS: Critically endangered

◎

Along with the turtle, efforts were made in the Chambal to reintroduce gharials, and met with considerable success. The gharial, also known as the gavial, or the fish-eating crocodile, gets its popular name from a bulbous projection shaped like a *ghara* or mud pot at the end of its snout. The Chambal sanctuary has the largest contiguous demographic of this species, reportedly between 48 per cent and 85 per cent of the global population. That's why the sanctuary, home to one of the first captive-breeding programmes for increasing wild gharial populations, is virtually the last stand for this odd-looking reptile.

It was in 1975 that the government of India set up Project Crocodile with the support of the United Nations Development Programme and the Food and Agriculture Organization of the United Nations. Project Crocodile set to work with a sense of urgency for the benefit of the three endangered Indian crocodilians—the mugger, the saltwater crocodile, and the gharial. It delineated 20,000 square kilometres as sanctuary land and set up several captive-breeding projects.

Renowned herpetologist Romulus Whitaker, who was instrumental in setting up the Madras Crocodile Bank Trust, noted that '[over] the next 10 years [since its inception] [that the Project Crocodile was in action, over 12,000 gharial eggs were collected from the wild and captive bred nests and over 5,000 gharials reared to about a metre or more in length and released in the wild. Over 3,500 of these were released in the Chambal River alone, which is the biggest of all the protected areas for gharials at over 425 river kilometers in length. Researchers like S.A. Hussain of the Wildlife Institute of India, R.J. Rao of Jiwaji University and R.K. Sharma of the Madhya Pradesh Forest Department spent weeks, every year, surveying, studying and counting along the Chambal River'.

However, after an energetic start, Project Crocodile started to wind down. Many conservationists blame the non-involvement of local people in the conservation of river resources, which restricted the scheme from becoming popular. Though committed researchers in the state wildlife departments, the Wildlife Institute of India and others continued to carry out surveys as and when they could or as part of research projects, not enough was done by the government to ensure the survival of the gharial even in its protected habitat. Damming, sand mining, and commercial fishing with gill nets ensured that the gharial was facing its deadliest period in 100 million years of existence. Not surprisingly, in 2004, R.K. Sharma, a gharial biologist from the MP Forest Department, found that gharial numbers had nosedived and there was visible degradation of the habitat, and communicated this to the sanctuary authorities.

Two years later, in 2006, in my capacity as a reporter for a news channel, I visited stretches of the Chambal because of reports that illegal sand mining had ravaged the sanctuary. I had the good fortune to tour the river with stalwart scientist Dhrubajyoti Basu, who had pioneered the initial studies in the region on gharials. As we set up our cameras by the river, we encountered a battery of men at work loading a row of trucks with sand from the sanctuary. There was literally no one to stop this daylight robbery of natural resources from a sanctuary. Sand mining was debilitating; and what was worse, it was undoing all the progress the captive-breeding programme had seen.

And then in 2007 a further blow was delivered: The gharials had hit the headlines for a tragic reason. The UP Forest Department confirmed that over 100 gharials had died in the Chambal in a mass die-off. No one knew the reason for these mass deaths. The media images were alarming—dead gharials with bloated stomachs floating in the waters. The MP, UP, and Rajasthan state governments pressed the panic button. The WWF-India was called upon for help, as were crocodile experts from across the world. An expert committee consisting of forest officers from all three states as well as scientists like Romulus Whitaker was formed. The postmortem showed that the gharials had died of gout, a disease of the kidneys. Of course, there were many unanswered questions: If the river had been contaminated by some toxic substance, why were only gharials dying? Why weren't other species like the mugger or the birds or turtles being affected? And why only young and sub-adult gharials?

A joint study taken up by the Turtle Survival Alliance and the Madras Crocodile Bank Trust, with support from the ministry of Environment and Forests and Climate Change, confirmed a high accumulation of certain toxins in tilapia. The report stated: 'It is believed that Tilapia—an introduced fish species—could have been carrying the toxins and excessive consumption of the same by gharials caused these mortalities.' Through the years, gharials have migrated upstream from the Chambal into the Yamuna during the monsoon. Either they had ingested toxins during the migration or the fish they fed on were contaminated. There was evidence to corroborate the latter theory. Mass fish die-offs in the Yamuna had been reported on at least three occasions in the seven–eight months prior to the gharial deaths. However, Jeff Lang, known for his extensive research on the Chambal, is not convinced. In an email interview, he states: 'The conclusion that tainted tilapia from the Yamuna caused the die-off is a "just so" story, plain and simple. The Yamuna has been badly polluted for decades, and also the toxin loads in tilapia are suspected but not well confirmed. More importantly, our studies have shown that the younger gharials (from hatchlings through the subadult stage) are real "homebodies", rarely venturing more than a few [kilometres from where they reside at that age], not even moving up and down the river during the monsoonal floods. Most subadults barely traverse more than 15–20 kilometres from where they live, year round. The epicentre of the die-off was just below Udi at Khera Ajab Singh, some 50+ kilometres upriver from the Chambal–Yamuna confluence. Only the adults move long distances to the confluence of the Yamuna every year to feed heavily on tilapia now, but few adults died during the mass die-off.'

There are still many gaps in our knowledge of what actually happened in 2007; fortunately, no further mass die-offs of the gharial have occurred since. If anything, the incident proved that even after years of scientific research and a captive-breeding programme, there was still so much that was unknown about the animals.

In order to fill in the gaps in information, small studies have been taken up by fitting the gharials with radio transmitters. This will provide vital information, for instance, on what happens to the gharial hatchlings after they are released in the river. Perhaps, the most fascinating study is being led by Jeff in the Chambal. His fieldwork has shown that reproductive or nesting females travel long distances. Explains Jeff via email: 'Our current longest traveler, [who] moves back and forth on a seasonal basis, is a female. She traveled upstream to a nesting sandbank at a turning in the river, 120 km from where she was tagged. She nested there, guarded her nest, opened it, and stayed with a large creche of 400–500 babies (from 8 to 10 nearby nests) until mid-July, then went 210 km downstream to the confluence as soon as monsoon waters flooded nesting banks, and the babies all dispersed. In November, she returned

210 km upstream to where she had earlier nested, and continued to do this each year for three more years, as long as she could be tracked.' Studies like these provide a peep into the world of these creatures, about whom so little is known. They also help ensure the success of the rewilding programme, as the gharials depend on the continuity of the river as their 'expressway' for survival.

Another study was initiated by Lang, in partnership with WWF-India, to monitor the long-term survival of gharials that had been released by the latter from the gharial-breeding facility in Kukrail into the Ganga at the Hastinapur Sanctuary in UP. The species had, at one point, been completely wiped out from India's most sacred river. Between 2009 and 2013, WWF-India released more than 100 gharials here. Again, the monitoring of reptilians after their release was poor. Writes Lang in an email: 'I have been working with their biologist, Sanjeev Yadav of WWF and we (with our crew from the Chambal) caught some of these released animals late last March (in 2015). However, more follow-up studies are needed on the part of WWF to determine what happens to the gharials in the unprotected parts of the sanctuary. These have simply not been done.' Despite this, the reintroduction in the upper Ganga, where gharials were once abundant, is groundbreaking, he feels, because it demonstrates that given the right conditions, gharials can be re-established and, hopefully, survive to build a self-sustaining wild population. The reintroduction of the gharials in the Ganga is important also from a symbolic point of view as the reptile gets its biological name from the river.

Sadly for the species, a number of external factors have undermined the conservation effort. India's growth story will have an adverse impact on the future of the gharials, especially when it comes to illegal sand mining, which is rampant in the Chambal sanctuary. While some efforts have been made by the forest department to curb it, strict enforcement continues to be a problem. India has the world's third-largest construction business after China and the US. Malls, houses, offices, and flyovers have sprung up all over, and infrastructure projects are booming, increasing the demand for sand in construction. A report in *Down to Earth* states: 'Legal or illegal, sand continues to be a scarce commodity. The construction sector, mostly real estate, constantly complains of acute shortage of this minor mineral.' On an average, one person uses 200 kilograms of sand per year, state reports by the Centre for Techno-Economic Mineral Policy Options (C-Tempo), a society registered under the ministry of Mines.

That's why Rom stresses that 'captive breeding of gharials is not needed anymore; what is needed are scientific assessments of the health of rivers and doing something about the dismal assessments of over-extraction of water, dams, diversions, and, of course, pollution.' This view is path-breaking, as Rom was one of the pioneers of captive breeding for crocodilian species. What Rom is now urging is a shift in policy that focuses on their habitat instead of the gharials themselves.

Lang is scathing in his criticism of the captive-breeding programmes. 'The incomparable wildlife of the Chambal simply needs an intact and open flowing river, with natural riverside habitats left intact, major curtailing of water extraction and sand mining, and to a lesser extent illegal fishing, poaching of turtles, etc. Left alone, with the natural ingredients in place, the wildlife will thrive.'

Today, the gharial's domain is a mere 2 per cent of its former range, limited to a couple of hundred square kilometres, and dwindling. The future of the gharial is so threatened that its IUCN Red List status was recently revised from endangered to critically endangered, one stop away from extinction. It is, today, the most endangered large animal in India, more so than the tiger. Given that the threats to gharials have never been addressed, nor the existing conflicts mitigated, it makes little sense to keep releasing thousands of hapless young gharials into rivers.

"Despite the enormity of past failures, reintroductions have neither stopped nor been critically evaluated. On the contrary, the pressure to allow such arbitrary releases is high even today because of captive-breeding successes, the resultant overcrowding in zoos and rearing centres, and the 'feel-good' factor, advises Rom. The reintroductions into the Chambal need to be followed up with better monitoring of the released animals. For without better monitoring procedures in place, such efforts result in millions of rupees down the drain without serving any conservation purpose. And, in the bargain, further threaten the lives of animals who were endangered to begin with.

5
Once I Caught a Fish Alive

◎

COMMON NAME: Big-headed fish [English]; Mahseer [Hindi]
SCIENTIFIC NAME: Golden Mahseer (Tor putitora), Deccan
Mahseer (Tor khudree), Humpback Mahseer (*Tor remadevii*)
CONSERVATION STATUS: Ranging from Endangered to Threatened

◎

The research for this book took me to India's far corners in search of odd creatures. And that's how I ended up driving hundreds of kilometres in search of the country's aquatic tiger—the mahseer. But for its slimy skin, the mahseer is a beautiful fish. And that explains why there is an entire sport dedicated to catching this tirelessly aggressive fish. What sparked my interest in this case was the involvement of a private agency in a rewilding project—and for a species that was not perceived as a charismatic one.

Fish as a taxon have been largely ignored in Indian conservation policies despite the fact that India is rich in aquatic biodiversity. Of the 39,000 species of fish recorded in the world, 2,500 are found in this country. The mahseer, a type of carp, is a hugely popular game fish endemic to the rain-fed rivers of Asia. They inhabit pristine freshwaters usually equated with untamed wilderness, and are, therefore, considered indicators of healthy freshwater ecosystems. Out of the 17 occurring in India, 13 have been assessed by IUCN for their conservation status. Of these, five species are considered near threatened, one critically endangered, three endangered, while there isn't enough data for the other four species.

First identified by Scottish physician Francis Buchanan Hamilton in 1822, the fish has been mentioned in several books and records of the 19th century. It has been favourably compared to other notable game fish, including the salmon, the tarpon (this, by none other than Rudyard Kipling), and the trout. With its large head, distinct fold of flesh below the lower lip, bright scales, and variously coloured fins, the mahseer is an attractive fish. It is also a fierce fighter—anglers who have wrestled with it talk about broken rods and snapped lines. All these qualities make it a favourite with hunters. To make matters worse, there is no legal framework to protect the mahseer, which explains why most species today are endangered. More recently, in January 2014, the Bombay Natural History Society in its magazine, *Hornbill*, published an article titled 'Where have all the Mahaseer gone?' in which the authors stated: 'Though mahaseer contribute in a major way to inland fisheries, they are one of the most threatened groups in the mega fishes. Unsustainable fishing and dams are some of the major threats to the mahaseer.'

A decline in mahseer numbers is not a recent phenomenon: Concerned about the decline of the mahseer in Indian rivers and reservoirs, S. Moolgaonkar, the late managing director of Tata Motors, and S.P. Manaktala, former managing director of Tata Power, approached C.V. Kulkarni, the director of fisheries for Maharashtra, in 1970, to support the rehabilitation of the species. In 1971, Shashank N. Ogale (now retired) joined Tata Power; he had a postgraduate degree in zoology and fisheries science. Spearheading the company's mahseer project, he began breeding the fish artificially. A 1976 government of India report by the National Commission on Agriculture also emphasized the need for mahseer conservation.

The fish were procured from a reservoir where the Tatas had constructed a dam and a hatchery was established in Lonavala for their breeding. The first batch of approximately 14,000 eggs was procured through a process known as dry stripping (a method of artificial reproduction in which the eggs from the female fish are stripped into a container). These were artificially fertilized and 10,000 were brought to maturity. Since then, the mahseer project has gathered the momentum required to sustain itself. Today, the hatchery has the capacity to hatch over 5 lakh eggs at a time. On an average, Tata Power breeds over 1 to 1.5 lakh mahseer fry each year and gives them gratis to many states in India. (Baby fish are initially called larvae, then fry, and finally fingerlings before they mature into adult fish.) Till 2017, it had produced more than 1 crore fingerlings of mahseer and distributed them across the country. Its findings on breeding methods have proved valuable and have even been replicated in several mahseer-breeding facilities across the country.

So how and why did a private company get into mahseer conservation? The story begins with the construction of a dam. For a hundred years now, Walwan Dam has stood across the Indrayani, not far from Lonavala. The dam was set up after Jamshedji Tata envisioned clean energy powering industry, particularly the cotton mills of

Mumbai. That dream was eventually realized by his son Dorab Tata, who laid the foundation stone of the dam on 8 February 1911. Today, the dam stands 26.63 metres high (above the lowest foundation) and 1,356 metres long, with a storage capacity of 72 million cubic metres, now boosted from the original 40 MW. Then, electricity was primarily transmitted from Khopoli in the Raigad district of Maharashtra to Parel in Mumbai to run the cotton mills. Now, it is also used by households in Lonavala and villages nearby. It was the power plant that led to the Tatas' involvement with captive breeding of the mahseer. 'The construction of dams in the river valleys had a lot to do with its [the mahseer's] decline. Mahseer breeds by migrating upstream. Structures, such as dams, began to block this process. There were also instances where the fish were killed in large numbers by detonating dynamite underwater for commercial fishing,' said Shashank, who spearheaded Tata Power's mahseer project. Vivek Talwar, who was the head of Tata Power's Sustainability Mission and is now retired, explained in an email: 'When we learnt about a species of freshwater fish that was once abundant in the Indrayani river and its tributaries, facing multiple man-made threats, including dams (such as our very own Walwan), habitat loss, habitat degradation and over-fishing, the decision to work on the conservation of the mahseer was an obvious next step.'

The breeding period for the mahseer is during the monsoon, in July and August, as the species require a lot of dissolved oxygen to breed. Brooders—parent fish—are caught from the specially made ponds for the brooders' stock during the season, and fed twice a day with a groundnut cake and rice bran mixture. 'In any conservation project, it is important to feed the fish with natural food, and supplement it with nutritive food, especially during the breeding season,' explains Vivek Vishwasrao, head of Biodiversity at Tata Power. The Lonavala hatchery has evolved the 'flow-through' method, which involves keeping the eggs in flowing water to simulate conditions in the wild. Eggs are obtained from an adult female aged three years and above and are kept in a cement tank in eight trays, with each tray able to accommodate 30,000 eggs. The tank has a direct water-sprinkling system that ensures the water is kept oxygenated, which is necessary to bring the eggs to maturity. The eggs are checked daily—healthy ones remain transparent while foul ones turn opaque—and foul eggs are individually removed using a dropper. The dexterity with which the hatchery staff do this shows that they have honed their skills over the years. I watch as the eggs, no bigger than tiny pearls, glisten in the water. Much like a diamond picker who evaluates the shine and cut of the stone to ensure the accuracy of angles and positions of cuts or bores using a magnifying glass, the staff identify and check the progress of each egg.

Eggs hatch into 1–5-centimetre-long fries in three to four days. They are kept at the hatchery until they are about three to nine months, and then despatched across the country depending on the orders placed. So far, at least 12 state governments have requested for mahseer from the Tata facility as well as educational institutions like the Fisheries College at Ratnagiri and Mangalore. The remaining fish are released into the six lakes near the hatchery to repopulate the mahseer in the wild. I am given a brief opportunity to observe this tiger of the river waters up close—'mahseer' is a corruption of the Hindi 'maha' and 'sher', great tiger. I fold my sleeves and step into the pond where the adult mahseer are kept. The mahseer feels slimy, like soap that keeps slipping from my hands. As the fish flaps about in my hand in protest, I can understand why it's been described as 'one of the fiercest fighting freshwater game fish with unparalleled strength and endurance'. I ease it back into the pond, thinking it's a maha sher indeed.

The Lonavala hatchery has come up with many innovations, including the way they despatch the eggs. Ogale and Kulkarni discuss this method in a paper in the *Journal of Aquaculture*. Mahseer eggs are tough, heavily laden with yolk, and take about 80 hours to hatch. As part of this experiment, 5,000 fertilized eggs, hardened for 24 hours, were packed in moist cotton wool and placed in a plastic basket within an outer container, then brought to Mumbai, a 100 kilometres away. They were sent unattended as air cargo to Bengaluru, where they arrived in the evening. On the first occasion, mortality on arrival at Bengaluru airport was 8 per cent, but in the second trial, with the same number of eggs, it reduced to 1.5 per cent. The eggs later hatched with 75 per cent survival at the Karnataka government's Hesaragetta fish farm. Ogale notes in the paper: 'This is probably the first case of air transport of fish eggs of Indian species in moist cotton wool. The success indicates good possibilities of dispatching live eggs of mahseer to distant places both within India and abroad, if required.'

Even as the scientists at the Lonavala hatchery were able to fine-tune their science, demands for mahseer fingerlings were coming from across India. However, there are very poor records of what happened to the fish once they were released into the wild, including their survival rates. Did they even make it back into the wild or were they simply kept in captive fish farms? Though information is available about the state governments and private agencies that requested Tata Power for the captive-bred fish, there seems to be no record of what happened to them thereafter.

This is a major lacuna of the project: that it did nothing to change the condition of the mahseer in the wild or address some of the conservation concerns that wiped out the species from rivers across India to begin with. That is why, despite a captive-breeding project that's lasted for more than four decades, the species continues to figure on the IUCN Red List. A breeding programme, however well-meaning, that does not tie into any conservation goals for the species in the wild will have only limited success. The goal of any rewilding project must be to improve the state of the species in the wild.

A seminal paper by Adrian Pinder and Rajeev Raghavan in May 2015 critiqued the efforts made by Tata Power and proved to be a game changer in mahseer conservation (Pinder is with the University of Bournemouth, while Raghavan is with the Kerala Univesity of Fisheries and Ocean Studies; both are members of the Mahseer Trust, which was set up to 'conserve mahseer as flagship species, to draw awareness and greater environmental stewardship to rivers'.). The authors focused their study on the river Cauvery in Karnataka and argued that the river contained two species of Mahseer, 'blue-finned' and an orange-finned or 'hump-backed' fish. Published in an international research journal, the paper showed that the endemic orange-finned mahseer was on the brink of extinction, having been replaced by non-native relatives (the blue-finned mahseer). It suggested that a Tata Power project that led to the 'reintroduction' of the non-native blue-finned mahseer in the Cauvery may have wiped out the latter. They went on to argue that this iconic fish was now swimming towards extinction.

The authors had been studying the ecology, taxonomy, and conservation status of the 17 species of mahseer which populate rivers throughout South and Southeast Asia from 2010 onwards. They looked at the angler-catch data in the Cauvery and came to realize that the river's mahseer community had undergone considerable shifts in the last 30 years.

The scholars argued that the Cauvery had blue-finned and hump-backed mahseer. 'Whilst it is not yet known whether these are distinct species or two different phenotypes, evidence suggests that the hump-backed phenotype is endemic to the river, whereas the blue-finned phenotype was introduced in the 1980s. Angler-catch data from a managed fishery on the River Cauvery, gathered between 1998 and 2012 and comprising 23,620 hrs. of fishing effort, revealed that captured individuals ranged in size from 0.45 to 46.8 kg, with the blue-finned phenotype comprising 95 per cent of all captured fish and the remainder being hump-backed.' The research suggested that the introduction of non-native mahseer acted as a catalyst which has had a catastrophic effect on the numbers of endemic hump-backed mahseer remaining in the Cauvery and its tributaries. 'In 1998, the ratio of hump-backed to blue-finned mahseer was 1:4. By 2012, this had dropped to 1:218,' says Pinder in his paper. Even worse, the researchers established that the hump-backed mahseer population was declining alarmingly, even as the blue-finned mahseer was slowly taking over the habitat. The latter, they found, 'originated from the TEC (Tata Electric Company) hatchery and was introduced during the late 1980s'. Tata Power, when contacted, maintains that it has always worked with credible agencies and experts for the programme; the company's role has been that of the supplier of the fish and that it did not reintroduce the fish.

Nonetheless, the paper challenged the assumption that reintroduction of the mahseer had contributed to conservation of the species. Pinder also wrote pieces in the popular press, sharing the findings of his study with the larger public. In an article in *Sanctuary Asia* magazine, he wrote about the experimental hybridization of mahseer species under Tata's Mahseer Conservation Project and the export of fingerlings throughout India and to Laos: 'While the pioneering work and considerable skill required to produce these fish in captivity is commendable, the impact of stocking several hundred thousand fingerlings into the Cauvery (from 1975 to present date [2015]) without first establishing a baseline of the endemic mahseer and an appreciation of their ecology has now been shown to have been implicated in the collapse of the hump-backed population which is today sadly spiraling towards extinction.' Pinder substantiated this point further in an email: 'We presented this very point to Tata's chief of sustainability (Vivek Talwar) and his team. This was met with agreement to the point we have since held a workshop with WWFI [WWF-India]/ BNHS [Bombay Natural History Society] and other major stakeholders and concluded that to date the Tata efforts have represented a breeding programme and not a conservation programme. This is now being corrected by Tata supporting Mahseer Trust work to try and re-establish the hump-backed mahseer within its endemic range. They have also said that they will stop the supply of blue fin mahseer.'

For his part, Vivek Talwar confirmed Pinder's point in an email, albeit with a rider: 'We have conducted an internal exercise to scrutinize research papers authored by Pinder and Raghavan on the mahseer as well as reached out to relevant expert organizations and people in the Indian conservation landscape, such as Dr A.J.T. Johnsingh and Mr Ravi Singh from WWF who is a member of Tata Power's Sustainability Advisory

Council, to get second opinions on this matter. Based on our assessments, we concluded that the case presented by Pinder on the blue-fins being the reason for the decline of the humpback mahseer in the Cauvery is a very good case of *correlation*, but not that of *causation*. Perhaps we may never be certain of the real cause of the decline of the humpback. This said and done, due to the kind of ethos the company nurtures at its core, we believe in preserving the environment and its components irrespective of what the cause of the decline might be.'

Pinder and Raghavan's findings had clearly stirred a hornet's nest. On the positive side, Pinder's email—as well as Talwar's—show that after the paper was published, Tata was open to course-correct, and in the next phase of the project, efforts are on to focus attention on the hump-back mahseer and save it from extinction. In 2018, Tata Power launched a programme for conserving the endangered humpback mahseer through an awareness drive with school children in the region.

Differences aside, a conservation plan certainly is the need of the hour, and Pinder acknowledges that now there is an effort to synergize and work with Tata Power and to do something to save the hump-back mahseer in the Cauvery. Again, Talwar is candid: 'This is one of the first instances where a private company stepped forward for the conservation of an endangered species. We have agreed to join hands with Pinder and work *together* instead of in opposition. In fact, the humpback conservation project in partnership with the Mahseer Trust is the next step in our journey of maturity of the mahaseer program and is slated to begin this year itself. This also ties in perfectly with our objective of expanding our programs so that we go beyond our neighborhood, and create a positive impact beyond [it].'

On the heels of the controversy of the mahseer breeding programme came another hotly contested debate—whether mahseer fishing by anglers for sport was helping or destroying the species. In India, angling as a sport is old as the British Empire. People had long fished India's rivers, streams, ponds, and seas for food and trade, but it was the British who popularized angling—fishing for sport—as a pastime. Coincidentally, the birth of India's wildlife parks can be traced to an angling expedition. While fishing near the Ramganga River with Sir Malcolm Halley in 1936, Jim Corbett, then governor of the erstwhile United Provinces, formalized the idea of setting up India's first national park. Though famous as a tiger hunter and later as a conservationist, Corbett was also a big angler. His fishing exploits with the mahseer, a feisty fish he described as the 'tiger of the water', became the subject of his epic tale, *Fish of My Dreams*.

With anglers from across the globe flocking to India's rivers, specialized angling outfits used to offer fishing expeditions across the country. Magazines specializing in angling were replete with advertisements from tour companies with the following promise: Chase Corbett's legacy as you tackle the Golden Mahseer in stunning locales, catch Brown and Rainbow Trout in Himalayan foothills, take on the monster Goonch or Giant Catfish on the Saryu (Mahakali), grapple with a Humpback Mahseer in the Cauvery, or go marine fishing off the coast of India for some Tuna, Barracuda and Giant Trevally. India was, clearly, a tourist destination for angling.

The Wildlife Association of South India (WASI), an NGO based in Bengaluru, Karnataka, came into existence in 1972 with a mandate 'to conserve and preserve the wildlife of South India'. The association also obtained a lease of a 22-kilometre-stretch of the Cauvery with the aim to conserve native mahseer populations. The focus of WASI's effort was to control illegal fishing and replenish wild species using captive-bred fish. The NGO also set up small seasonal fishing camps to promote responsible 'catch and release' mahseer fisheries. Their success encouraged other NGOs, such as the Coorg Wildlife Society and the state government—owned Jungle Lodges and Resorts to set up both seasonal and full-time angling camps on the Cauvery during the 1980s and 1990s. A similar initiative was also set up in north India in the Jim Corbett National Park. The income generated from recreational fisheries effectively controlled the illegal fishing of mahseer through the establishment of anti-poaching camps, as well as the rehabilitation of former poachers as 'ghillies' or fishing guides, thus providing them alternative employment and its associated societal benefits.

Despite the effective participatory conservation model practised on the Cauvery, on 17 April 2009, a legal notice was issued under Section 55 of the Indian Wildlife (Protection) Act. It questioned the construction (albeit temporary) of the privately owned Bush Betta fishing camp within the Cauvery Wildlife Sanctuary, without prior approval from the National Wildlife Board and the Supreme Court. This was followed by the issue of a further legal notice to the Central Empowerment Committee of the Supreme Court, drawing attention to the further violation of the Wildlife Protection Act by permitting angling within the boundaries of the Cauvery Wildlife Sanctuary. Doubts were also raised about whether the catch-and-release method was damaging the mahseer in any way or affected its ability to survive once released back in the river.

In a paper published in the journal *Fisheries Management and Ecology* titled 'Rapid Assessment of the Physiological Impact Caused by Catch and Release Angling of the Blue Finned Mahseer in the Cauvery River', the findings suggested that overall, injuries were found to be minor in nature and mortality from to the sport—which involved catching the fish through a hook that penetrated the mouth—was negligible. Throughout the study, only one fish was considered likely to die, but no cases of mortality were observed during the study period. Of the 44 angled blue-finned mahseer assessed for hooking location, most (91 per cent) were hooked in the mouth, specifically in the corner of the mouth. Four fish (9 per cent) were foul-hooked, and each instance of foul-hooking was also categorized as a minor, moderate, or major injury, according to the degree of resulting tissue damage. The paper concluded: 'The rapid assessment findings suggest mahseer are robust to C&R [catch and release], but also provide data to support the development of best angling practices designed to reduce unnecessarily long angling times and air exposures.'

Pinder has also argued in favour of angling in different research papers: 'Apart from the positive role played by recreational fishing, the success of these efforts also demonstrated the importance of engaging local communities in the conservation of endemic and threatened freshwater fish species. Recreational fishers constitute a social group that offers unique potential to enhance fish conservation. They have a vested interest in preserving or enhancing the resources they depend on and there is ample evidence to demonstrate that anglers work proactively to conserve, and where possible enhance, aquatic biodiversity.' Despite strong evidence that showed that angling as a sport could generate local livelihoods and help in the conservation of the mahseer, the temporary camps on the Cauvery and the Ramganga in Corbett were asked to shut shop, as they defied sections of the Wildlife Protection Act.

The involvement of the private sector in the conservation of the mahseer has shown that long-term commitment can bring some, if not all, desirable outputs. That Tata Power has managed to keep the mahseer population intact in their area of operation and streamlined captive-breeding protocols for this fish are some of the major achievements. However, the lack of data on fish released into the wild are gaps in information that the company must address. The most important lesson learnt is the willingness of a private company to correct course and link its breeding programme to a conservation goal through partnership with experts who could guide them. It now remains to be seen how this unique partnership between the Mahseer Trust, an

NGO, and Tata Power will pan out. If it works, it will certainly be a step forward for the hump-back mahseer, which will hopefully be seen more abundantly in the Cauvery waters.

The larger question of what is happening to the state of our rivers—the waters that the mahseer rides—is beyond the realm of Tata Power's breeding programme. The onus cannot be on one private player to save this species. The wild habitat of the mahseer is under severe threat from dams, pollution, and over-fishing. This is where state governments need to step in. Could the way forward be through angling, which could provide incentives to local communities? Could the religious and cultural values associated with the fish (for instance at places like Rishikesh where pilgrims feed schools of mahseer fish in the Ganga) be also drawn on, to propagate its protection? These are points to ponder on, so that the fish of Jim Corbett's dreams become abundant in the wild.

6

The Vultures Have Landed

◎

COMMON NAME: Oriental white-backed vulture, Long-billed
vulture, Slender-billed vulture [English]; Geedh [Hindi]
SCIENTIFIC NAME: *Gyps bengalensis, Gyps indicus, Gyps tenuirostris*
CONSERVATION STATUS: Critically Endangered

◎

Vultures have a special place in my heart. My first close encounter with one nudged me in the direction of environment journalism. This was in my days as a school student. I was bending over to drink some water from the water cooler when I heard a loud thud: a black object had landed next to my feet. I realized it was a bird, an Indian Gyps vulture I'd learn later, its throat almost slit in half. On closer examination,

I found the *manjha* (string) of a kite had got entangled around its wings, hugging it in an embrace of death.

This was in August, when kite-flying is at its peak in Delhi. I watched the bird struggle to get up, only to collapse again, all the while hissing loudly as a warning to keep away. I'd been rescuing injured dogs and cats for a while now, and taking them to a nearby animal shelter, but this gigantic bird posed a challenge. I asked around for help. Finally, some members of a TV crew that happened to be shooting in school that day helped me lift and wrap the heavy bird in a cloth to take it to an animal shelter. The bird finally passed away at the shelter, unable to recover from its injuries.

The incident left an indelible mark on me. I went on to write about it for a national daily. At the age of 16, I did not know then that this brief incident would set me off on a lifelong pursuit of writing about wild animals, the forests they inhabit, and the perils they face. That article was about an injured individual; I didn't know then that within a decade the entire species would vanish from our skies.

The crisis facing our vultures is well documented—a deadly toxin nearly knocked them off the biological map, with populations crashing across South Asia. In India, a few individuals were taken in captivity with the possibility of breeding them in large numbers so as to repopulate them in the wild. While much has been written about the captive-breeding centres, what remains untold are the unconventional methods employed to save them: conservationists have doubled up as detectives, lobbied with the pharmaceutical industry, and ingeniously replicated their natural habitat in their breeding enclosures.

It was in pursuit of such stories that I travelled to the foothills of the Shivaliks. On a small patch of land on the edge of the city of Chandigarh, conservation history is in the making. At the Jatayu Conservation Breeding Centre in Pinjore, Haryana, a handful of conservation biologists have captive-bred 250 birds that were on their way down the extinction vortex. An international consortium of scientists and NGOs from USA, UK, Pakistan, Nepal, and India came together to grant these birds a new lease of life, by not just breeding them in captivity but lobbying with various governments to get the deadly toxin that caused their near wipeout banned. Seventeen years after the first breeding centre was set up in 2001, the first batch of birds was ready to make it to the wild.

Ornithologist Vibhu Prakash (who now heads Jatayu) is among those who have seen life come full circle for vultures. He and renowned bird expert Dr Asad Rahmani (who went on to become the director of the Bombay Natural History Society in 1997) were among the first people to ring the alarm about crashing vulture populations.

Vibhu was doing research at the world-famous bird park in Bharatpur, at the Keoladeo National Park in Rajasthan, when he noticed something was amiss. Between 1985 and 1997, the population of the oriental white-backed vulture (*Gyps bengalensis*) had declined by an estimated 97 per cent at Keoladeo; by 2003, this colony had become extinct.

Estimates suggest that in the 1980s, the populations of vultures (including white-backed, long-billed, and slender-billed) were roughly around 40 million or 4 crore, but this number plummeted to a few thousands in the decade to follow. In 1996, an animal keeper in Lucknow made an innocuous statement to Rahmani about how difficult it had become to spot vultures. This appeared in Rahmani's cover story ('Race to Save the Vultures') on the birds in BNHS's *Hornbill* magazine. That observation fortuitously sparked off India's biggest bird rescue project, and it soon extended across the subcontinent. Following feedback from several scientists and field workers in different parts of the country, a series of surveys was carried out to confirm initial suspicions. In 2000, the BNHS teams undertook over 11,000 kilometres of road-based surveys, repeating

6,000 kilometres of road transects previously surveyed for raptors in the early 1990s. They confirmed that declines of more than 92 per cent had occurred in all regions across northern India. The organization repeated surveys covering the same route and methodology in 2001, 2003, and 2007 to monitor trends in numbers.

The results of the intensive research sent shockwaves through the conservation community. Of the nine species of vultures found in India, four were now listed as Critically Endangered and one as Endangered. Catastrophic decline in populations of three species of the resident Gyps vultures (*oriental white-backed vulture Gyps bengalensis, Long-billed vulture G. indicus* and *Slender-billed vulture G. tenuirostris*) were observed.

But the surveys showed rapid declines; India's oriental white-backed vultures declined at an average rate of 48 per cent a year (an aggregate of 11,000 birds) for the period from 2001 to 2007. Long-billed (45,000) and slender-billed (1,000) vultures were estimated to have declined at around 22 per cent a year. Populations of red-headed vultures and Egyptian vultures had also declined, at the rate of 41 per cent and 35 per cent a year in India. The survey in 2007 indicated that numbers of oriental white-backed vultures had declined by a staggering 99.9 per cent over the preceding 15 years. Long-billed and slender-billed vultures decreased by

97 per cent over the same period. Surveys across Nepal and Pakistan indicated vultures had declined at similar rates across South Asia. In fact, in Pakistan, both resident species (white-backed and long-billed) were on the edge of extinction.

South Asia was once a haven for these birds because of the large amount of 'food' available to them, in the form of the sheer number of cattle reared across the subcontinent. (In India and Nepal, cows are considered sacred for Hindus, and are not eaten, so their carcasses were the principal food source for resident species.) Government statistics indicate that livestock numbers in India exceeded 40 crore since the 1980s and rose to over 50 crore in 2005. Parsis in India and Buddhist communities on the Tibetan plateau traditionally used these birds for sky burials so as to cleanly and efficiently dispose of their dead.

The cause of the rapid plunge in the population was traced to diclofenac, a non-steriodal, anti-inflammatory drug (NSAID) lethal to vultures. Since vultures constituted the natural animal-disposal system, it seemed obvious that there was something in the carcass that was causing the death. The studies carried out by BNHS and international organizations found the usual suspects—pesticide poisoning, industrial pollutants, or food shortage—did not show anything abnormal in the vultures. First, an infectious disease agent was suspected.

A two-year investigation by an international team of 13 scientists led by Lindsey Oak of Peregrine Fund of USA finally identified diclofenac as the likely culprit. The drug was used by farmers and veterinarians to ease pain in cattle. An overdose of the drug can cause kidney damage in humans, and it seemed to be a likely cause of death in the vultures. Further tests established that there were residues of diclofenac in dead vultures.

Once the cause was confirmed, the lobbying with the government and pharmaceutical companies began to get the deadly drug banned. Nita Shah, a biologist who had earlier worked in the Rann of Kutch, was appointed as the advocacy officer by BNHS to liaise with the government and several ministries (like health and environment) to make this happen. Rahmani says that once the government realized the gravity of the situation, it was 'one of the fastest decisions taken' for a country known for its red-tapism. It was also praiseworthy, given that governments in other parts of the world, especially Europe, had taken over 30 years to ban DDT, the killer pesticide in the late 1960s.

It was in 2006, after a series of meetings with the government, that the vultures had their first victory. The National Board for Wildlife in India, based on scientific evidence, recommended a ban on the veterinary use of diclofenac on 17 March. In May 2006, a directive from the Drug Controller General of India was circulated, requiring the withdrawal of manufacturing licences for veterinary formulations of diclofenac. This directive was further strengthened in 2008, when it was made a punishable offence to manufacture, retail, or use diclofenac for veterinary purposes. So how effective was the ban on the drug? Once again, data helped clarify if policy measures had been effective. A study conducted by BNHS and other organizations, such as the Royal Society for Protection of Birds, in 2011 found that five years after the ban, there was a perceptible change in the veterinary use of the drug. Data was collected from three surveys: the first was in May 2004–July 2005, before the ban; the second in April–December 2006, immediately after; and the third was in January 2007–December 2008, a few years after the ban came into effect. Liver samples were taken from a large number of sites distributed across the northern half of India, predominantly from cattle and water buffalo carcasses deposited at carcass dumps managed by local government corporations, co-operatives, private companies, individuals, and cattle welfare charities. Samples were also collected from slaughterhouses. Sampling locations were typical of sites formerly used by large numbers of foraging Gyps vultures. In the seven to 31 months after the first implementation of the ban, the prevalence and concentration of diclofenac in carcasses of domesticated ungulates that served as food for vultures had almost been halved. More importantly, following the ban, the rate of decline of vulture populations had also gone down, indicating that it had been an effective conservation tool. An estimate of the rate of decline in Oriental white-backed vulture populations in India, from road-transect counts, showed an annual decline of 44 per cent between 2000 and 2007. After the ban, the new estimated annual decline rate was 18 per cent. The figures showed that conservationists had won the first step of the battle. The next step lay in releasing the captive vultures into the wild.

At the Jatayu centre, Vibhu stares at six screens of CCTV footage with the keenness of a security officer. For Vibhu, this is more than just treasure—it is what he has devoted the best years of his life to: closely observing the 200-odd birds they have managed to breed in captivity. When the vulture crisis was detected and a decision was taken to breed them in captivity, Vibhu and his wife Nikita moved from the bustle of Mumbai to a small town close to the Jatayu centre in Pinjore to turn the fate of the vultures around. Nikita, herself a scientist, was finishing her doctoral research on vultures and was appointed as the vulture egg incubation specialist.

A number of CCTV cameras are installed in the large aviaries at the Jatayu centre, helping Vibhu and his team to observe the birds' behaviour, their breeding activities, and their feeding patterns, all without having to expose them to human contact; for successful reintroduction into the wild, it is imperative that the birds do not get used to humans being around. The centre's nursery provides a nest-like environment for the chicks, and there's a hospital for sick and injured birds. The vultures are fed on the meat of goats housed under the centre's management and is checked to ensure there is no diclofenac. Perches have been positioned strategically to recreate a natural environment. The large aviaries allow the birds to exercise their wings over a significant area. The hope is that these scavengers, after breeding to significantly large numbers, will return to the wild. In January 2010, under the watchful care of Nikita and Vibhu, the first artificially incubated egg was hatched. That triggered hope that the vultures could now be saved, even though it was merely the start of a very long journey.

Vibhu applied all the practical knowledge he had learnt from studying vultures in the wild at this facility. During his time in Bharatpur, he had made a small machan from a charpoy, where he'd sit for hours to study raptors. Months later, when he went back to dismantle the charpoy, he found that vultures were using his ingenious machan to roost. He used this observation from the field to design perches for the vultures at Jatayu. And that's how you can see several of these light bedsteads strung with coir suspended rather awkwardly from the centre's walls. For the resident vultures, they provide just the right kind of support and comfort to roost.

When this adaptation proved a success, Vibhu experimented some more: he lined the wooden stands with coir to recreate a branch of the tree on which a vulture would perch in the wild. This again provided a cushion for the vultures in captivity. The cushion was important since they perched for longer hours in captivity than they would probably in the wild. 'The birds have taught us everything. Once an egg cracked but did not hatch for couple of days. We noted that the male bird swooped down and took a dip in the bowls of water we keep in the aviary, then flew back to his nest and sat on the eggs. The eggs hatched within a couple of hours. We learnt then that high humidity was important for vulture eggs to hatch. There is so much we don't know and we are still learning every day,' he confesses.

Another of the innovations at the centre is a method called 'double clutching', which helps optimize resources. A vulture lays only one egg at a time, but if that egg is removed from its nest within 10–15 days of laying, the confused mother will lay another egg. The first one is put into an artificial incubator and monitored round the clock till it hatches. The egg is rolled around from time to time, just as the mother and father would do in the wild. Nikita admits that this is like bringing up a baby—perhaps tougher. Artificial incubation is commonly used on poultry farms, but the stakes there are much lower. 'Here,' says Nikita, 'we just can't afford to lose an egg.' To compound matters, vultures are also slow breeders—a pair can take years to hatch a healthy baby.

Another problem of breeding birds in captivity is ensuring a balanced male–female ratio. The problem with vultures is that they are not sexually dimorphic, that is, they don't display distinct sexually characteristic behaviour. Their sex, therefore, is determined using their DNA. Nikita elaborates, 'We are aware of the sex of most of the adults, but a number of juveniles are still unsexed. There are facilities available at the centre to sex the birds, and this coming summer, we will attempt to do so for all the birds. We will exchange birds between various centres to main equal sex ratios. Vultures pair for life. This is very important so that we do not waste colony aviary space. This would also be the ideal time to exchange birds within the three older centres to make sure that we have the optimum blood lines and no potential inbreeding.'

While learning about the birds and watching them on the CCTVs is engaging, there is nothing quite like observing them for real. I silently walk outside, to a corner of the woods where a small hide overlooks the large aviaries. I can see the grand vultures perched high up on the ledge. Images of the bird I helped rescue as a child, and the odd-looking bile it had produced, flash past my mind's eye. I describe it to Vibhu, who tells me that it is a strategy to ward off predators—vomitting and producing strong-smelling bile. And he is right; I recall how the injured bird had produced that bile each time I had tried to get closer to rescue it. I peek at the birds from my spot, watching in silence, hoping that one day, these birds will soar high in the sky.

After the success of the Jatayu centre, the BNHS established two more centres to fulfil the objective of releasing 600 pairs of each of the threatened species to form a genetically viable and self-sustaining population. The Central Zoo Authority then sanctioned four more centres in the country in different states with the help of the zoos to help achieve the objective of the conservation breeding programme. The Bhopal centre was given to BNHS for day-to-day running. In all, the governments of India, Pakistan, and Nepal set up 10 centres for breeding vultures in captivity.

Even as Vibhu and his team worked hard on breeding the birds and training them for the wild in captivity, there was the challenge of removing the toxins from the environment completely. As Rahmani noted in a special issue of *Mistnet* (a quarterly newsletter of the Indian Bird Conservation Network) that was devoted entirely to the concept of 'vulture safe zones' or VSZ: 'One of the crucial components of any conservation breeding and release programme is the removal of the cause(s) that made the species rare in the first instance and this is stated as a precondition in the IUCN guidelines for reintroduction. If the main cause(s) is not removed, the captive-bred individuals will die. In any case, the mortality of captive-bred animals after release is from 40 to 60 per cent, as some unfortunately cannot adapt themselves to living in the wild.' Rahmani went on to warn, 'If the main reasons for rarity are still prevalent, there could be 100 per cent mortality among the released individuals. The whole purpose of the conservation breeding programme would fail.' The main objective of conservation breeding is insurance against extinction. The birds would not be released unless the cause of mortality was removed from the system.

This is why the BNHS-RSPB (Royal Society for Protection of Birds) partnership that was spearheading the vulture conservation work now shifted focus to develop VSZs covering 30,000–40,000 square kilometres each, where captive-bred vultures could be released. The idea was that it would be easier to retain the small but key remaining vulture populations in the wild through VSZs, where there would be a very low risk of poisoning. These sites will be vitally important in the future, not just for the numbers they retain within a natural system, but because they are also likely to be utilized as some of the first release sites for captive-reared birds. Release efforts will be focused in areas where it has been established that vultures can be protected and birds are likely to congregate.

Among all the vulture conservation programmes initiated in South Asia, Bird Conservation Nepal was the first to pioneer the concept. In fact, in Nepal, VSZs and 'vulture restaurants', also known as 'Jatayu' restaurants, have been combined successfully as conservation tools. At these sites, old cattle are bought or donated and cared for until their natural death. Only meloxicam (the safe alternative to diclofenac) or ayurvedic medicines are used for pain relief. Diclofenac (if still used, despite the ban) can remain in cattle for approximately one week; therefore, as a precaution, only carcasses of cattle that have been kept for a minimum of 10 days are put out for vultures and the others are buried. Any Jatayu restaurant in South Asia that does not adhere to this simple rule is likely to provide vultures contaminated food. The final element of the programme is to attract vultures to the safe area and to retain those that are already there by providing them safe food regularly. To ensure safe food, cow shelters have been established in the villages surrounding the vulture colonies. These shelters buy old cattle that are at the end of their working lives and are otherwise destined to be sold to cattle traders for use as meat—or else abandoned by their owners in forest land or outside villages. In Nepal, old cattle can be purchased for around $2 and many animals have been given to the project, as it saves local people from feeding or abandoning an animal that is otherwise a burden. The cattle are housed in purpose-built sheds and herded to fields on community-owned land in the village where they can graze. A project veterinarian ensures their welfare with regular checks and, if necessary, medical treatment (that excludes diclofenac use). No cattle are killed, but when the animals die, their carcasses are skinned (providing important income to the project to pay the cattle herder and purchase more old animals) and then placed out for vultures to feed upon. Flocks of over 150 vultures are now regularly seen at three in-situ conservation sites in Nepal. In India, the state forest department of Maharashtra set up a vulture restaurant in Phansad Wildlife Sanctuary, which offers diclofenac-free meat to vultures.

The creation of a VSZ is not easy and requires a number of intricate steps. This starts with defining the geographical area for implementation. Each provisional VSZ is centred on at least one surviving nesting colony of at least one of the three critically endangered Gyps vultures. The safe zone needs to cover an area that has a radius of at least 100 kilometres in every direction from the colony. This equals a total area of over 30,000 square kilometres. Such a large area may enclose hundreds of settlements, tens of millions of livestock, and may cross state and national borders. Once a VSZ has been identified, the next step is to assemble a team of field biologists and advocacy officers and build their capacity for advocating vulture conservation and raising awareness of the diclofenac problem among decision-makers and stakeholders. This covers a vast cross-section of people, from bureaucrats in the environment ministry to pharmaceutical companies, down to the district collector and ordinary farmers, and engaging with each of them for vulture conservation.

Undercover pharmacy surveys form a crucial step in the creation of the VSZ. Very simply, these surveys are done with a person (trained by the team) acting as a cattle owner and going to pharmacies that sell veterinary drugs. They would then ask for an injectable painkiller for treating the injured cattle for pain and inflammation.

The first drug offered is purchased by the surveyor. Such surveys are done across the entire VSZ area. In Pinjore, such a survey was carried out between 2015 to 2017 and found that the prevalence of diclofenac decreased from 50 per cent to 20 per cent in pharmacies. This was positive feedback for Vibhu and his team. The pharma surveys are an important tool to evaluate the possibility of toxins in the VSZ. When the surveys stop finding diclofenac and other vulture-toxic NSAIDs on sale, only then can plausible steps be considered for releasing the captive-bred vultures. At least six other drugs—aceclofenac, carprofen, flunixin, ketoprofen, nimesulide, and phenylbutazone—have been found to be toxic for vultures.

While India has taken concrete measures to ban diclofenac, it is still readily available across the counter in many European countries. The killer drug was approved by the Spanish Agency for Medicines and Health Products in March 2013, in spite of the fact that Spain is home to 90 per cent of Europe's wild vultures. The European Medicines Agency confirmed that veterinary diclofenac does present a clear and present danger to European vultures, and recommended that a number of risk-management measures be taken to avoid the chance of their being poisoned. These included more regulation, veterinary controls, better labelling, better information availability, and/or a ban of the drug. Despite the fact that wildlife groups in Europe, such as the SEO BirdLife (Spain), the Portuguese Society for the Study of Birds, Vulture Conservation Foundation, BirdLife Europe, and World Wide Fund for Nature, launched a sustained campaign for the drug's ban across the continent, it continues to be available. It is ironic that developing economies in South Asia took the extreme step to ban a drug and committed funds for the birds' captive breeding, but developed European countries have been dragging their feet on the matter despite conclusive scientific evidence.

It's been a long and arduous journey for the vulture—and for those who have been battling to keep it alive. Vibhu recalls moments of frustration: 'At first, some local journalists kept trying to find fault with our work here.' That explains why he is wary of talking to anyone from the media. 'There has been irresponsible journalism sometimes, but we also need the media to create awareness about what we do, so I realize I have to engage with them,' he admits. There is always the constant challenge of new variants of the toxic drug becoming available, so long-term pressure on pharma companies has to continue. On the positive side, many battles have indeed been won: All vulture-range states in the Indian subcontinent have banned the veterinary use of diclofenac; regular monitoring of residues in cattle carcasses shows that the level of diclofenac contamination of the vulture food supply has fallen substantially. Safety testing has identified meloxicam as a safe alternative drug; in fact, monitoring NSAID residues in cattle carcasses shows that its use has increased markedly in India, and it has become widely available in Nepal and Bangladesh as well. All of this points to an optimistic future—of seeing vultures dot our skies again one day.

As I pull away from the Jatayu centre, I notice the splurge of construction at the foothills of the Shivaliks—shopping malls, glitzy resorts, and a slew of development projects eating away like termites at our wild places, wiping out yet another patch of forest. Yet, Vibhu's words of hope echo in my ears: 'The vulture is a resilient creature. It will survive. Yes, we need the forests for species like the leopard or the tiger, but the good news is the vulture can manage to survive even outside of the forest, *provided* we are able to remove the deadly toxins from its environment.' Surely, the vultures have landed, I think.

With less than 150 individuals in the wild, the pygmy hog, the smallest member of the pig family is considered critically endangered.

Source: Vijay Bedi. Reproduced with permission.

The Mahseer are a group of species of freshwater fish, most of which face the threat of extinction. A private company is breeding them in captivity but will that be enough to restore their numbers in the wild?

Source: Adrian Pinder. Reproduced with permission.

Lines of captive elephants were used to flush out wild hogs into a net where they were captured to establish the first breeding centre for pygmy hogs in India.

Source: Goutam Narayan. Reproduced with permission.

The Red-crowned Roofed Turtle (*Batagur kachuga*) is a critically endangered freshwater turtle species. Efforts are on to repopulate the species in the wild through riverside hatcheries and better habitat protection.

Source: Vijay Bedi. Reproduced with permission.

Panna National Park in Madhya Pradesh was once home to over 40 tigers that were lost to poaching. Efforts are on to bring back the big cat by releasing tigers from other areas here.

Source: Joanna Van Gruisen. Reproduced with permission.

Launched in 2005, Indian **Rhino** Vision **2020** is an ambitious effort to attain a wild population of at least 3,000 greater **one-horned rhinos** spread over seven protected areas in the Indian state of Assam by the year **2020**.

Source: Vijay Bedi. Reproduced with permission.

The Jatayu Conservation Breeding Centre in Haryana is the world's largest facility for breeding vultures in captivity. Here the mighty birds feed outside a pre-release aviary at the Centre.

Source: Vibhu Prakash Mathur and Nikita Prakash. Reproduced with permission.

7

The Return of the Unicorns

◎

COMMON NAME: Indian one-horned rhinoceros [English];
Gainda [Hindi]; Gor [Assamese]
SCIENTIFIC NAME: *Rhinoceros unicornis*
CONSERVATION STATUS: Vulnerable

◎

Imagine encountering a steely one-horned rhinoceros on your evening walk in the park. Pobitora, a small wildlife sanctuary in Northeast India, on the floodplains of the grand Brahmaputra, provides the opportunity in spades. The grey 'unicorns', grazing so casually, are found in every nook and corner here, making it hard to reconcile to the fact that there's less than 4,000 of them left in the wild. There's a mother with a calf who snorts at us angrily, raising her heavy, one-horned nose in the air as trigger-happy photographers surround her. A few kilometres ahead, we spot a male rhino with some bruise marks on its armour plates, the result, perhaps, of a fight for territory. The Pobitora Wildlife Sanctuary is no bigger than 39 square kilometres, yet it is home to over 100 one-horned rhinos—and they're jostling for space with wild buffaloes, barking deer, and over thousands of migratory waterfowl that arrive here in winter. Pobitora may be full up—but in this problem of plenty lies an opportunity.

The situation has brought about a conservation coup in Assam in the form of the grand Indian Rhino Vision (IRV) 2020, a coordinated conservation policy that has seen scientists and NGOs band with the government. Under this vision, animals are being shifted from protected areas, such as Pobitora and Kaziranga, where they abound, to those like Manas, where the twin pressures of insurgency and poaching have wiped out local populations. Ungulates like the Swamp deer (*Rucervus duvaucelii*) are also part of this grand plan to rewild parks in Assam. It's been a gamble, given the losses along the way, but now, hopefully, the vision's on its way to becoming a grand success.

Assam is, in fact, the proud host of more than two-thirds of the global population of the one-horned rhino, even though the animal is a frequent victim of poaching. Topping the revival story is the World Heritage Site of Manas; its fate bends and folds quite like the Manasa River from which it gets its name. The park almost lost this special status bestowed on it by UNESCO after a period of internal strife wiped out its entire rhino population along with those of animals like tigers and deer. However, after a decade of hard work, efforts to rewild Manas are finally yielding results. A ceasefire with the extremist group that once ruled the park has helped the forest staff record significant conservation gains.

In my days as a broadcast journalist reporting on wildlife, I got to see countless video clips sent in by reporters from across the country—from leopards being chased by a mob in rural Andhra Pradesh to a tiger being born in a zoo—every picture was worth a thousand stories. But one video clip that's etched in my mind was from Manas, of a baby rhino standing helplessly next to the body of its mother, who had bled to death after its horn had been sawed off mercilessly by poachers. The baby was dehydrated and traumatized, refusing to leave the mother's mutilated body. This was in 2013, when a rhino that had been reintroduced to the park was shot down. It showed the desperation of the poachers (who had sawed off her horn even though she wasn't yet dead), the misery of the female rhino tortured in her final moments, and the reality of an illegal trade that stretches across international borders. As I stood there in the newsroom watching the live feed come in from a remote part of Assam, I knew this was a story that had to be told.

Now, a few years later and for the purpose of this book, I am on my maiden visit to Manas under happier circumstances: there has been a lull in poaching in the last few years. Wildlife biologist Amit Sharma, from WWF-India, is accompanying me. Amit was at the front line of the reintroduction effort in Manas, which saw 18 rhinos brought in. 'Poaching in Manas is cyclical,' he tells me. 'It comes and goes in waves, so we can never say it is finished.' Right now, though, the world heritage site is riding the peace wave. In 2017, its famous green hero, Bibhuti Lakhdar, had been given the IUCN award for his efforts to restore this landscape. Assam's most famous forest staff, D.D. Boro, the hero of many landmark wildlife documentaries, transferred here from Kaziranga, has been appointed here as an officer on special duty. Boro flashes his famous wide grin and says he is happy to be back among his people. His smile reflects an optimism that is shared by many, from the WWF staff to the forest guards. It is, however, an optimism that is laced with caution.

It's a sentiment echoed by park director H.K. Sharma, who knows it may be too soon to call Manas a conservation success story—the threat of poaching is always circling around the park, daring them. Sharma operates under several constraints. Many posts in the forest department are lying empty, with no one interested in joining the wildlife wing, he tells me. I ask him if the security of a government job doesn't serve as enough temptation, to which he retorts, 'Who wants to leave their families behind and live in a war zone?' He may be right; Manas has a long, chequered history, not just of poaching but of insurgency as well. Even now, there are whispers that parts of Manas are not really within the control of the forest department. Some of the guards mention in informal chats that they are too scared to venture into the western range of the park, which is under the control of insurgents. And that's not the only problem that Sharma faces: in spite of all the international talk of saving tigers and rhinos and the glory that this park has received over the years, the forest staff has not been paid in eight months. 'My men are patrolling on the ground with no salaries,' he says. That, it seems, is how conservation operates in the real world.

Given the ground reality, it's clear that IRV 2020 is an ambitious programme—the objective being to attain a population of 3,000 wild rhinos across Assam, distributed over seven of its protected areas by the year 2020. Begun in 2008, it was decided to halt the translocation by 2013 because of the rampant poaching of translocated rhinos. In 2013, just as conservationists were celebrating the birth of two calves born to reintroduced rhinos, a heart-rending piece of news came in. One of these mothers had been poached for her horn, leaving behind a two-week-old calf. The calf was rescued and hand-reared. So why are conservationists refusing to give up on Manas, engulfed as it is by episodes of poaching and insurgency? It is to understand this that I find myself in the park.

I spend several days in Manas moving around in a noisy SUV, searching for the 'Big 5'—rhinos, tigers, wild buffaloes, gaurs, and elephants—as we make our way through the tall grass. I am equally interested in smaller species; Manas is after all the last stronghold of the wild populations of the pygmy hog, the smallest pig in the world, but their sightings are indeed rare. Quite often, I request the forest guard to let us just switch off the cantankerous vehicle to take in the sounds of the forest. 'Come in March,' suggests Amit, as I strain my eyes to catch a glimpse of perhaps a spotted deer. 'That's when the forest staff sets fire to the grass to create fire breaks.' These gaps or 'breaks' in vegetation or other combustible material act as a barrier—brakes, if you will—to slow or stop the progress of wildfires. 'It's easier to view wildlife then as the tall grass is cleared.' After an hour of driving around the park, we climb up a watch tower in the park, and a herd of elephants comes forward and surrounds us. The matriarch stands at a distance as the young ones coil their trunks around a clump of grass or indulge in a bout of play. The herd is in no hurry; it stays at the watchtower for over an hour pulling at the grass gingerly, snapping branches off the nearby trees, providing entertainment to a vehicle full of tourists under the ever-alert presence of a forest guard with a rifle. As the last rays of the sun dance on the yellow grass, the giant greys melt into the forest as silently as they had arrived, allowing the relieved tourists who'd been biding their time to make their exit.

At another watchtower, we hear reports of the sighting of the lesser Bengal florican, a bird known for its unique mating ritual of jumping up and down like a giggly toddler on a trampoline. I squint through the binoculars, itching to catch a glimpse of this endangered bird, but only manage to see a white blur in the distance. On our drive back to the forest rest house, we see a nightjar squatting stubbornly on the road like an agitating protestor, a male wild boar scurrying off into the bush on its short legs almost as if it forgot something at home, and a mighty gaur with a tuft of grass hanging from its mouth like a hungry tourist stuffing his face with spaghetti. The actors of this animal theatre have much to do before the sun sets and the nocturnal actors take centre stage. We are staying inside the park today, at the Mathangudi Forest Rest House, with the happiest country in the world, Bhutan, on the other side of the river Manasa.

At night, under the dim yellow light, my attempts at updating my daily notes are hindered by a howling wind that creeps in through the windows, making my hands tremble as I struggle to keep warm. The blankets are musty and overused from the swarms of picnicking tourists who come here. Having survived the cold night, I welcome the warmth of the sun the next morning, and after a breakfast of warm rotis and chai, prepare myself for another glorious day in the park. We drive around the forest at a slow pace, encountering wildlife, stopping at lonely forest watchtowers, chatting with the front-line staff engaged with the onerous task of protecting the wild inhabitants of Manas, and gulping down black tea laced with lots of sugar. Fresh milk is a rarity. It's a minor hardship though, compared to the life-threatening struggles the staff deal with daily. The threat of possible gunfire, patrolling among tall grasses that decrease visibility—these are only part of the physical hardships they endure. There is the psychological hardship of being away from their families for weeks on end, not to mention living in the isolation of the watchtowers. It's the story of the hundreds of countless Indian soldiers as well who patrol our borders, except in this case, not many people are aware of these soldiers of our forests.

The story of Manas National Park would be incomplete without understanding the political struggles of the Bodo people, recognized as one of the largest hill tribes under the 6th Schedule of the Indian Constitution. The conservation successes or failures of the park have been incumbent on the demand by the Bodos for a separate state. Their struggle for self-determination originated in the colonial period: as early as the 1930s, Gurudev Kalicharan Brahma, the then lone leader of the Bodos, submitted a memorandum to the Simon Commission for a separate political administration for the indigenous and tribal people of Assam. However, the British ignored his demand.

Even after the country gained independence, this struggle continued, leading to a volatile mix of politics and ethnic conflict in the region. In 1987, the Bodo people raised a demand for a separate state consisting of areas located in the extreme north, on the northern bank of the Brahmaputra. The six-year-long vigorous Bodoland movement culminated in the first tripartite Bodo Accord on 20 February 1993 with the All Bodo

Students Union (ABSU), the central government, and the Assam government as signatories. This paved the way for the creation of the erstwhile Bodoland Autonomous Council (BAC) and the suspension of the demand for a separate state. The ABSU revived the movement in 1996, claiming that the BAC had failed to fulfil the aspirations of the Bodos.

Meanwhile, the erstwhile militant outfit, the Bodo Liberation Tigers (BLT), launched an armed struggle for statehood, leading to further incidents of violence. On 10 February 2003, the second tripartite Bodo Accord was signed by the BLT with the centre and state governments. The Bodoland Territorial Council (BTC) was created under an amended provision of the 6th Schedule, which led to ABSU suspending the movement. Manas is currently under the control of the BTC and the forest department comes second to it as far as managing the park is concerned. This is unlike other parts of the country where the state forest department has the primary responsibility for managing the protected area. That's why the story of conservation in Manas is embedded in struggle for identity of the Bodo people. Even as Manas was being repopulated with rhinos, NGOs like the Wildlife Trust of India (WTI) were reaching out to the BTC for conservation gains. In August 2016, a grand announcement was made that would boost the ongoing rewilding efforts in the region. BTC's deputy chief Kampa Borgoyary announced that an additional 350 square kilometres would be added to the park's existing 500 square kilometres, increasing the total area to 850 square kilometres and creating more space for rhinos and tigers. This was the first step to creating a larger landscape for wildlife.

Following the ceasefire agreement in 2002 and the creation of the BTC, some semblance of peace returned to Manas. The time was ripe to undo the damage of the past and look at reintroducing the wildlife species that had been lost in the intervening years of strife—all its rhinos, and, it is believed, almost 50 per cent of its tiger species. In February 2006, for the first time in India, a hand-raised rhino calf was translocated, by the Assam forest department and the WTI to Manas. WTI was already working in Kaziranga, rescuing wild animals that got injured or separated from their mothers during the annual floods. And this rhino rehabilitation project endeavoured to give orphaned or displaced hand-raised rhinos from Kaziranga a home in Manas.

I got an opportunity to visit the rescue centre, the Centre for Wildlife Rehabilitation and Conservation (CWRC), at Kaziranga, which had rescued three rhino babies that had been orphaned when floods hit the park in 2016. At the centre, a team of doctors and keepers would monitor the orphans' health over the course of a few months, until they were old enough to be released into the wild. I watched them from a window that opened into a small, open grass enclosure where they were being kept, nudging each other playfully. As their human foster mothers arrived with giant milk bottles to feed them, the trauma of separation from their biological mothers seemed to be forgotten, at least momentarily.

Over the years, as part of this project, WTI has released eight rhinos in Manas. The process that is followed is simple: after their translocation from CWRC, the rhinos are released into a spacious African-style semi-open enclosure, called *boma*, spanning about 33 acres that has been created in Manas. The rhinos will be confined here until they attain sexual maturity. The boma ensures that the calves are safe from predators even as it helps them acclimatize to the local environment. These rhinos have no interaction with humans, except during periodic medical assessments. After about two or three years of acclimatization, the calves are released into the wild and are remotely monitored round the clock with the help of radio transmitters.

In addition to this, the government of Assam, with support from the BTC and NGOS like WWF-India and the International Rhino Foundation, got involved in reviving Manas through the implementation of IRV 2020; the first wild-to-wild rhino translocations began in April 2008 and continued in phases till 2012. In all, 18 Indian rhinos were moved from Pobitora and Kaziranga to Manas. Eleven rhino calves have been born in the park since then.

Unfortunately, this population has also been tragically touched by poaching: there was a region-wide increase in poaching in 2012–13. After losing eight animals, IRV 2020 halted rhino translocations to Manas in 2013 to focus on improving security. Training in new patrolling methods, along with the support of new park leadership, made a big difference—only one rhino was lost to poaching in Manas in 2014. The current population of 33 rhinos and new calves continues to thrive.

But the story of Manas is not just about saving the rhino. The park is one of the very few places that have grasslands, the habitat of the last-surviving populations of endangered species: floricans, rhinos, pygmy hogs, and the hispid hare. And so, rewilding experiments need to focus not just on reintroducing endangered animals, but also on saving their habitat. For a forest department that has been trained since the colonial period to plant fast-growing trees, this has meant reorienting their mindset. Restoring the ecology in this part of the world, therefore, meant uprooting trees to bring back grassland species while educating the guardians of these forests about the necessity to do so. This was easier said than done, given that one-fifth of the grasslands in Manas have been invaded by weeds. The rapidly colonizing alien plants found here—*Mikania micrantha* (Mikania) and *Chromolaena odorata* (Chromolaena)—are among the world's most invasive plants. These South American species are fast growing and capable of producing large amounts of biomass, and were introduced into India early last century either for ornamental or agricultural purposes. Both plants also secrete chemicals in the soil that inhibit the growth of other plants. That's why IRV 2020 also focused on reviving the grasslands in Manas and combating the invasive species choking the habitat. It's also why the IUCN insists that before an animal is reintroduced in the wild, it's essential that its habitat be saved.

IRV 2020 also looked at species other than rhinos— the eastern swamp deer (*Rucervus duvaucelii ranjitsinhi*), for instance, a sub-species of the swamp deer found in eastern India. According to a scientific observation in *Oryx* in 2013, some photographic evidence of the presence of swamp deer in Manas was captured, but organizations like WWF admitted that further research was needed. The deer, formerly distributed all over the Brahmaputra floodplains and the Terai foothills of the

Eastern Himalayas, are now found in a single isolated population in Kaziranga. With less than a thousand individuals, this population is facing an extinction threat from various anthropogenic as well as biological forces. This deer sub-species has always remained in the shadow of the more charismatic rhino, tiger, and elephant, and thus been neglected. That oversight was one of the main reasons it was believed to have been wiped out from Manas and confined to a small population in Kaziranga. In 2014, as part of the effort to reintroduce faunal species in Manas, 19 eastern swamp deer were transferred from Kaziranga. A team of experts from the state forest department, the WTI, and the College of Veterinary Science from Guwahati monitored the entire process. The animals were released in a purpose-built boma secured by a two-line power fence to deter leopards from entering. The boma was also flooded, and short grass from nearby areas was transplanted to ensure the well-being of the herd. According to WTI, the translocation has been successful, and perhaps the best indicator has been the birth of a fawn in 2014, signalling that the animals had settled in their new habitat. Since then, wildlife biologist Rathin Burman from WTI, who co-ordinated these efforts, has observed a population of 25 individuals of two herds located in Kuribeel (14 individuals) and Bhatghali (11 individuals) area of the

Bansbari range of Manas National Park. Both the areas are well protected and have suitable habitat for the deer to thrive. The efforts to reintroduce swamp deer would no doubt have also bolstered the existing population in Manas, of which some photographic evidence had been gathered in recent years.

Even as all these conservation efforts are on, what do the people who live around Manas think of the wild animals in their backyard? I drive to Palchiguri, less than a kilometre from the fringes of Manas. It is a picturesque village, with a pond and thatched, mud-washed huts in a neat row. I am here to meet village elder Bijoy Boro. In his late 60s, Boro has served as the head of the panchayat that includes three villages nearby. He offers me *tamul* (a combination of betel leaf with areca nut) as I sit in his courtyard to understand how his community views the efforts to bring back the one-horned rhino and other wild animals to Manas. Behind his thatched hut, the women of the house are separating the husk from the rice.

Boro talks about how the Bodo community traditionally considered a rhino sighting a blessing from the gods, Lakshmi in particular and said to bring luck. 'Seeing a rhino in the wild was a blessing, and we consider the urine and dung of the rhino as sacred.' In fact, if they found rhino dung, they'd collect it and dry it to treat skin diseases. But he jokes, 'These days there is too much good luck.' He is referring to the tendency of the hand-reared rhinos to wander into the village.

There are more than 50 other villages around the park. Ranging from fisheries to dairy farming to beekeeping and weaving, sporadic efforts have been made by different NGOs to work with the communities that live around Manas. However, what was unclear (as is the case with most community-based conservation projects that are imposed from outside) was *how* this created goodwill for conservation, and transformed the locals into advocates of conservation. As I visit one of the women's group in a village, the leader of the group asks me, 'So can we hope to get anything from you through this visit?' This tendency to view outsiders as offering free sops signifies all that is wrong with such programmes.

For instance, I observed that one NGO had distributed smokeless stoves in a village bordering Manas to stop people from collecting firewood from the park. Over time and with constant use, the stoves needed to be repaired or serviced, and since the women didn't know how to fix them, they stopped using them altogether. And so the people were back foraging in the forest for firewood. Could interventions have been made through a social forestry scheme for the locals to plant trees that could subsequently be used for collecting firewood? This would have had greater long-term impact—it would reduce the dependence on the national park for firewood and provide the benefits of a secondary forest that people could have access to. Such an effort could perhaps have generated goodwill for conservation rather than a technology that is alien to them.

Perhaps, the reason why the more long-term measures are never implemented is that they currently involve conservation biologists—essentially trained in the science of managing wild animals—to engage with communities. What is needed is for conservation groups to employ social scientists to work with the community and move beyond giving sops in the hope of promoting goodwill for conservation.

Another way to build bridges with the local people is to promote more community-based tourism in the area. This could create the necessary goodwill for conservation, provided it is done the right way—by empowering locals to manage and run the tourism initiatives rather than making them mere service providers. Some efforts have been made in this direction, such as the Manas Maozigendri Ecotourism Society (MMES), an ecotourism initiative set up through a memorandum of understanding between the village people and the forest department. Perhaps, the most important achievement of MMES has been to absorb former insurgents and poachers into its ranks as conservation volunteers. These people are now the torchbearers for the local community in ushering in peace and stability in the area. Today, the society is involved in ecotourism through a jungle tourist camp run solely by the community people. Another local community-tourism initiative known as MEWS is led by seven locals who got together to set up an accommodation facility for tourists. Gradually, over the years, the thatched huts have been lined with modern amenities and the staff have received extensive technical support from organizations like WWF-India.

The struggle to maintain and protect wildlife in Manas will always be a work in progress. Efforts are on to expand the scope of the park and thus create more space for biodiversity. The idea of the Transboundary Manas Conservation Area (TraMCA) was conceptualized in 2011 as a larger space that goes beyond national boundaries and connects protected areas, biological corridors, and adjoining reserve forests of southeastern Bhutan with that of northeastern India. The creation of TranMCA addresses the third 'C' that ecologists Michael Soulé and Reed Noss describe in the concept of rewilding, corridors; carnivores and cores being the other two. Another advantage would be that the landscape forms a vital mosaic of conservation spaces across the Eastern Himalayas. India's Manas National Park and Bhutan's Royal Manas National Park form the core of this space that would thus enlarge the homes of flagship species like tigers, elephants, rhinos, pygmy hog, Bengal florican, and hispid hare.

The royal government of Bhutan has already incorporated the Bhutan TraMCA component into their 11th Five Year Plan (2013–18). In fact, WWF-Bhutan is supporting the country's department of forests and park services to implement some of the key outcomes identified in the TraMCA Action Plan for Bhutan. What happens on the Indian side depends on the ability of the political establishment to maintain peace with the Bodos.

On my last day in Manas, I am blessed to witness one of nature's most dazzling spectacles, a ballet put up by Mother Nature against a light blue canvas, a swirling, pulsating, huge black cloud. This is the 'murmuration' of starlings moving in exquisite co-ordination, as if orchestrated by an unseen hand. Grainger Hunt, a scientist at the Peregrine Foundation, describes this phenomenon as an aerial ballet almost 'like a fluid choreography of funnels, ribbons, and hourglasses, spills and mixing, ever in motion. Dense in one moment, diffuse in the next.' Experts say it's basically a mass aerial stunt performed by birds before they roost for the night.

The flock of starlings provides this aerial display as a parting visual treat on my last evening here.

The swirling cloud of starlings is a reminder that Manas is once again rebounding with life. The 'unicorns' of Manas may have returned but will their continued safety be a pipe dream? The long-term survival of Manas's inhabitants is inextricably linked to the fate of the Bodo community. Once the political cauldron of conflict is addressed, conservation will be a piece of cake.

8
Of Pigeons and Paradoxes
Why urban rewilding matters

As a child, I spent hours in the mud patch outside my house watching the big black ants that arrived at the start of the monsoon walk in single file to the storm-water drain some distance away. I made tiny homes for them in the mud, decorating them with leaves and stones as I stared at the creatures in rapturous awe. I'm quite sure that I became a champion for green causes not because of any wildlife documentary I watched but because of close and regular interaction with nature in my formative years. Scholars who have researched the subject confirm this belief. In 1980, Thomas Tanner, from the University of Iowa, researched how 'significant life experiences' in early childhood that help engage with nature contribute to making environmentally conscious citizens.

But with biodiversity vanishing so fast from our cities, the opportunities to engage with nature are fewer and fewer. In spite of the fact that it's in cities that public policy around biodiversity and environmental protection is shaped, not too many rewilding projects have an urban base. 'Rewilding', for instance, is a term that is generally associated with reintroducing an apex predator into an ecosystem or repopulating a depleting species in a forest. But the tide is now turning—across the world, urban rewilding projects are becoming common. In Chicago, for instance, a not-for-profit called Urban Rivers manages a project called The Wild Mile, which is in the process of transforming the man-made, steel-walled North Branch Canal of the Chicago River into a haven for wildlife. Places like Seattle, Austin, and Salt Lake City are starting to experiment with installing permeable pavements in low-traffic areas, along with living infrastructure like green roofs and other vegetated areas planted with native species. Then there is the High Line Park project in New York City, which has a 2.3-kilometre-long elevated greenway that's created on a former railroad and is replete with natural grass and indigenous plant species.

It is estimated that over half of Earth's human population now lives in urban areas. Yet, there aren't enough incentives for conservation planners to focus their attention on saving biodiversity in cities, simply because of the belief that urban areas do not have large enough habitat blocks to sustain viable natural populations. Urbanization, therefore, presents fundamental challenges but also unprecedented opportunities to enhance the resilience and ecological functioning of urban systems.

At their most ambitious, urban restoration projects can actually succeed in managing rare species in such environments; at the very least, they can restore more diverse ecosystems than those represented by sidewalk cracks and abandoned parking lots.

But there is another offshoot of restoring ecosystems in urban areas. The engagement with nature can help create a support group for green causes. Children need first-hand experience with biodiversity to become passionate about its protection, and such projects can also provide opportunities for more active involvement, as with citizen science, restoration ecology, and environmental monitoring. Take, for example, the initiative to count the slender loris (*Loris tardigradus*) in Bengaluru. This small nocturnal primate, found only in Sri Lanka and southern India, has somehow survived in Bengaluru, a mushrooming megacity of 1 crore people. But a catastrophic loss of trees in what was formerly known as India's Garden City threatens their future. The Urban Slender Loris Project is a participatory collaborative study of urban biodiversity, bringing amateur citizens and professional scientists together to monitor areas where this primate is found. This initiative attempts to engage citizens from various walks of life in the science of conservation, as well as make them aware of the primate once found in their backyards. It is a great example of mobilizing people living in cities for the cause of conservation.

In this chapter, I set out to explore two examples of ecological restoration in urban settings, in Delhi and in Bengaluru. The first is an urban restoration/ rewilding project on a patch of forest taken up a group of citizens just on the edge of Delhi. The suburban city of Gurugram in north India, known for its high crime rate, massive traffic snarls, and snazzy high-rise corporate offices, symbolizes all that is going wrong with a city— poor planning, a depleting water table, and virtually no public transport. But an enthusiastic bunch of citizens have organized themselves under the banner of 'I AM GURGAON' to turn things around. Leading this group is 45-year-old Latika Thukral, a banker who gave up a corporate job to pursue her dream of setting up an NGO for a greener city—Gurugram, her home for the last 16 years. Over the past decade, the organization has succeeded in healing nearly 550 acres of land and turning it into the Aravali Biodiversity Park. Today, the park is a repository for native flora and fauna of the Aravalis; it has a water-conservation zone, an educational space to spread awareness about environmental issues—particularly among children—and a recreational space as well. In one visit to the park you can encounter a mixed coterie of people, from the morning walker to the avid birdwatcher.

The Aravalis once extended from Gurugram in Haryana all the way to Rajasthan and Gujarat, covering over 700 kilometres, and had their own unique biodiversity. Over the years, parts of the Aravalis, especially in Haryana, have been ravaged or lost due to rampant mining. Latika and her team decided to change all that. They could have just built an urban park for their morning walks. But they went a step further—and entered into a partnership with the government to build a biodiversity park. This meant planting indigenous trees and bringing back local species typical to the Aravalis. Now, with over 150–200 species of native trees, shrubs, herbs, climbers, and grasses; over 170 species of birds (black kite, shrike, drongo, sparrow and bulbul); and animals like the civet cat, jackal, nilgai, porcupine, hare, and mongoose, the biodiversity park is teeming with life. 'Our vision was to give Gurugram something it could cherish. Greenery was the answer since Gurugram is located in the Aravali ranges and used to be lush and green, before the hills were ravaged by extensive mining,' explains Latika. 'We took over the park in 2009 from the Haryana forest department, and since then, the survival rate of saplings has increased from 10 per cent to 40 per cent, which has encouraged us a lot.'

And encouragement was definitely the need of the hour: Latika and her friends faced many and varied roadblocks along the way, from administrative apathy to scarcity of water. Water was initially a big challenge. However, more than 40 companies in Gurugram provided them with adequate funds to nurture saplings. 'At present, we have two nurseries inside the park to grow saplings. We have one borewell inside the park and the rest we manage with treated water given to us by a local company,' she says. Plants must be selected carefully as random planting can do more harm than good. To remedy this, Latika planted trees that were once part of this ecosystem.

To many, the term rewilding or ecological restoration means planting trees or reviving a habitat. In this case, Latika and her team had to turn things on their head— and, rather ironically, begin by ridding the landscape of a profusion of vilayati kikar. This native Mexican tree species was introduced by the British in the 1800s to beautify the newly declared capital. The motive behind choosing this specific species was that it could adapt easily to the region's arid soil. However, the kikar caused great damage, wiping out local species and depleting the water table wherever it was planted. And that's how Latika and her team, advised by environmentalists, came to uproot this alien species before any actual work could be done.

This massive deforestation drive entailed working with the State forest department to change their mindset. The next step was finding people to donate saplings, to be planted over the 500 acres. That's where Latika's corporate background came in handy. She reached out to businesses with offices in Gurugram, who stepped up and donated in large numbers. The next step was finding a source of water. Though there was a borewell, they didn't want to exploit a depleting water table, as that would only add to their woes. So she made arrangements for drip irrigation, which had never been tried before for such a large area, and one which was undulating in parts. It worked. Now, they have an arrangement for using water from the Gurugram sewage treatment plant.

The Gurugram Municipal Corporation helped by fencing the entire park and solving many encroachment-related issues with locals from surrounding villages. Working with the government has had its ups and downs. While some officers have been co-operative, others had their own ideas of how to manage a park. Says Latika, 'One officer suggested having an open-air gymnasium, another wanted to build a night safari, and a third wanted to open a crocodile park. But by and large, we were able to bring them around to our concept of how we want to run this park.'

Renowned naturalist Vijay Dhasmana has been working with Latika and her team, spearheading the plantation drives, sourcing native species, and building the nursery, among other responsibilities. He believes that Latika and her team's work stands out for its emphasis on restoring ecosystem services and wanting to bring back the native species of the Aravalis. 'This is not just a recreational space,' he explains. 'The 550 acres are helping recharge a severely depleted ground water table for Gurugram. Moreover, we have helped bring back so many species that were once common to the Aravalis, such as *Boswellia serrata* (salai), *Sterculia urens* (kullu), *Mitragyna parvifolia* (kaim), *Lannea coromendelica* (gurjan), and *Anogeissus pendula* (dhau).

The citizen's initiative hasn't stopped just at restoring the Aravali Biodiversity Park. Propelled by the success of the park, they are now greening an abandoned storm-water drain or bundh that was once part of an ancient irrigation system built during the British Raj. Most of these bundhs or irrigation channels have become clogged with mountains of plastic, as they were being used largely as dumping grounds. Latika's team in 2017 removed 250 truckloads worth of plastic and debris from the drain.

The rejuvenation of the Chakkarpur–Wazirabad Bandh uses an integrated system for proper collection and discharge of rainwater. The drainage system includes recharge chambers, channels, trenches, pits, and perforated pipes to ensure that rainwater flows in the right direction and recharges the ground water table as well as the nullah. The project used 1,900 tractors of debris and 480 trucks of soil brought from various sites in and around Gurugram. Nearly 5,000 cubic metres of construction and demolition waste was used effectively, ensuring a reduction in dumping of roadside waste and efficient utilization of resources. The corridor has been restored by planting indigenous trees and developing a cycling track for the residents. Latika proudly states that a 1,020-metre stretch of bundh land has been transformed into a linear urban park in a span of one year.

The idea was to transform the bundh into a green corridor, much like The High Line Park in New York City has rejuvenated the disused Central Rail line into a 2.33-kilometre-long aerial greenway. Already, many pedestrians of all age groups use this stretch of the bundh daily, both as free public space and to walk across to the other side.

As with any urban green initiative, the success of the Aravali Biodiversity Park—as also the bundh project—rested on a number of critical facts. One, in the case of the biodiversity reserve, the unique partnership with the government allowed the NGO to focus on conservation aspects rather than get caught up in legal or infrastructural entanglements. The fact that the project had the backing of the government made interventions easier for them. Another is the involvement of local residents so that there is a sense of ownership over the project. That also explains the unprecedented popularity that the park enjoys among the citizens of Gurugram, who, in November 2018, showed up in the thousands to protest a proposal by the National Highway Authority of India (NHAI) to build an elevated road that would have wiped out the park. It was because of the citizens' protest that NHAI finally had to back off and come up with an alternate plan that would not destroy the green lung.

◎

Nearly 2,000 kilometres away from Gurugram, in India's Silicon Valley, a similar initiative was taken up.

In this case, the objective was the revival of a forgotten wetland.

Bengaluru's tree-lined avenues, bungalows with majestic gardens, large parks, and scenic lakes have, in the past few decades, faced an unprecedented onslaught. Rapid growth and urbanization have converted most of the wetlands into dumpyards or sites for discharge of industrial effluents. A study conducted by researchers from the Indian Institute of Science in March 2016 revealed only four of the 105 surveyed lakes in the city to be in good condition. Further, 25 lakes were in a very bad state, entirely covered with macrophytes (thus affecting the diversity of the lake) or dumped with solid or liquid waste and with little or no water. The most alarming finding was that not even one of these lakes had water that was fit for drinking according to standards laid down by the Central Pollution Control Board. Fish deaths and foam, indicating toxicity, were reported from most lakes across the city.

One of these wetlands, the Bellandur Lake, made international headlines when it routinely started spewing out foam and fire, creating the illusion of white snow spilling out onto the streets of Bengaluru, and creating a visual barrier for pedestrians and cyclists. This foam was a cocktail of toxic chemicals caused by the flow of untreated sewage water into the lakes, leading to an accumulation of high phosphorus and oil on the surface. The frothing and foaming Bellandur thus became a visual symbol of all that was going wrong with the city's lakes.

Even as Bellandur became a big issue, a group of citizens had decided to take matters into their own hands to transform another lake, the Kaikondrahalli, into a rich biodiverse ecosystem. Harini Nagendra, a professor of sustainability at the city's Azim Premji University, who has been closely associated with the revival efforts, outlines the need to examine the restoration in its historical context. 'Lakes were maintained by local communities who formed a closely intertwined social-ecological system, wherein certain communities took up responsibilities for various activities, such as channelling and monitoring water distribution, dredging and desilting the lake, or maintenance of the tank embankments. Water from the lakes was used for domestic purposes, agriculture, and to recharge large, open cast community wells placed next to these lakes, as well as for fishing.... Water occupies a significant position in Indian cultural and sacred traditions, and most lakes had lake deities next to the shore, which helped to reinforce restrictions on overuse during specific seasons, through cultural and religious taboos.' In the case of Bengaluru, these traditions were disrupted by the sheer power of rapid urbanization wiping out the intricate network of wetlands and lakes that the city had developed over centuries.

Spread over 48 acres, Kaikondrahalli Lake lies on the southeastern periphery of the city. In mid-2008, in response to a newspaper report that the Bengaluru Corporation was planning a restoration of this lake, local residents formed a group with experts in engineering, ecology, education, and outreach. This group would interface with the government in order to refine the plans, keeping in mind ecological principles. Priya Ramasubban, a documentary filmmaker who had moved to the area, was motivated to do something for the lake. A founding trustee of the Mahadevpura Parisara Samrakshane Mattu Abhivrudhi Samiti (MAPSAS) and an active advocate of the project, Priya says that she's always been drawn to nature, even as a child. She recalls that a good part of her childhood in Chennai was spent on turtle walks that a local NGO organized. When she moved to Bengaluru, she'd walk around the lake, and especially a section where there she'd seen hundreds of dragonflies swimming above the water. The moment, she recalls, was just 'mesmerizing'. It became a turning point for her association with the lake.

Gradually, more and more local residents got involved. The objective was to develop an integrated plan for lake restoration with a focus on conservation, water rejuvenation, and the socio-economic requirements of multiple strata of society. Unlike Gurugram, where the Aravali Biodiversity Park served a more homogenous crowd, the lake served a diverse group—from original inhabitants of peri-urban villages around the lake to occupants of newly built, high-end apartments and residences, sometimes with very different cultural and political beliefs. It took years of citizen participation, as well as partnering with the government, to transform the heavily polluted lake, receiving sewage from nearby residential areas, to its original state. The work included draining and dredging the basin to remove silt, diverting sewage, restoring the bundh, and planting a rich variety of locally adapted, biodiversity-friendly trees and plants around the periphery. The original restoration plans called for the conversion of some areas covered by water into landscaped gardens and the extension of the jogging path to the biodiversity-rich marshy area at one end of the lake. Based on input from the community, with additional advice from ornithologists and naturalists familiar with the lake and its non-human inhabitants, these plans were modified. Plans for an expensive, input-intensive landscaped garden was thankfully dropped, says Nagendra, and the walkway avoids the marshy parts that attract biodiversity.

The rejuvenation was done in two phases, between 2009 and 2011. Afterwards, some members of the core group came together to form MAPSAS. These included Priya, David Lewis, a retired senior citizen who contributed much of his time to overseeing daily activities, and Rajesh Rao, another local resident who oversaw the management of the neighbouring Ambalipura Lake. The trust entered into a tripartite agreement with the municipal body, the Bruhat Bengaluru Mahanagara Palike (BBMP), and a corporate funding body, United Way, to manage Kaikondrahalli Lake. Over the years, local residents have seen a *kere habba* (lake festival) held here. The first one organized in January 2015 attracted over 3,000 visitors in a single day, while by 2016 the numbers had swelled to 4,500 people. A mix of students from neighbourhood schools as well as from the adjacent slum participated in these festivals, painting stones and leaves and creating rangoli—temporary art installations with flowers and grasses. Economist and Nobel laureate Elinor Ostrom planted a tree near the lake in 2012 to honour the effort of the local residents in restoring this urban commons that is now popular with locals and visitors alike. A large and growing number of people living around the lake visit it frequently, and have participated in a number of activities associated with its restoration, maintenance, and fund-raising over the years. No entrance fees are charged and the lake is maintained using donations and funds from locals and organizations.

A year after restoration, the lake had attracted over 50 species of birds and a rich variety of butterflies, frogs, toads, and snakes. The variety of animal and insect biodiversity has grown substantially since then, with many more bird species added to this list. Biodiversity experts have recorded more than 43 species of birds, 26 species of reptiles, and 3,000 plants. Amateur naturalist and birder Vinodh Venugopal, who frequently visits the Kaikondrahalli Lake, has photographed different species depending on the season. He has seen pond herons, egrets, coots along with white-throated kingfishers, cormorants, and darters. In August, spot-billed pelicans, painted storks, and Asian open bills dominate for a couple of months. The two huge fig trees found here are home to large numbers of frugivorous birds like the Asian koel, white-cheeked barbet, rose-ringed parakeet, the common myna, and the jungle myna. Rat snakes and checkered keel backs are common during the monsoon, along with 11 species of frogs that have been identified here. The return of the birds, amphibians, and reptiles is an indicator that the lake ecology has been restored.

This idyllic scenario has not been achieved without a number of challenges, some of which are those of ongoing maintenance, while others include complex and persistent long-term issues. As Nagendra noted in a 2016 paper titled 'Restoration of the Kaikondrahalli Lake in Bangalore': 'A frequent problem is when neighbouring landowners and residents let in sewage into the lake, typically in the middle of the night when it is most difficult to monitor the lake. While the lake has been fortunate in having a large intact wetland upstream, which helps to clean up the sewage and recharge the lake in the monsoon, a number of buildings are slated to come up in this wetland in the coming years, which will severely impact the long-term sustainability of the lake. Despite the fact that this construction violates environmental norms, efforts by MAPSAS to halt construction have failed to make headway so far.'

Another problem has been finding regular funding for the project. While United Way came forward to finance the initial restoration work, there has been a struggle to find new sources of funding. Another challenge has been the dynamic nature of the citizens' involvement. The core set of people who were involved initially have moved on. Priya, for instance, has moved to Chiang Mai in Thailand, but in a telephone interview to me in 2019, she explains: 'I wanted to make sure that my involvement was such that after a point I became redundant. There were others who have subsequently joined with different skill sets who can raise funds or manage the administrative part of running a trust.' The dynamism of MAPSAS can also be seen in a positive way, as it ensures a fresh set of ideas coming in, with the baton being passed on to keep the restoration work sustainable.

The social challenge of lake rejuvenation in cities has been the difficulty of developing the impetus for collective action. The problems stem from high levels of diversity and inequity, constant change in the socioeconomic and cultural backdrops, and the apathy of real estate developers and many residents who, despite living in high-income neighbourhoods around these lakes and visiting them frequently, are not willing to contribute their time, effort, and/or money towards lake maintenance and funding.

While life has continued at Kaikondrahalli Lake despite the disturbances and challenges, it is a reminder once again that restoration efforts are a work in progress with new threats always round the corner. Collective action, observes Nagendra, 'is especially challenging in urban contexts, where people are on the move, preoccupied, and living in heterogeneous, dense, assemblages, often in relative anonymity. Using collective action to successfully facilitate the sustainable management of natural resources is even trickier in cities. There needs to be scope for the people affected by the mismanagement of urban forests and lakes (often the poorest of the urban poor) to be able to have some input into their management and monitoring, and prompt punishment of offenders'.

There are strong lessons to be learnt from these two case studies, not just in terms of organizing people but also for ecological restoration in urban settings. While Bengaluru was able to ensure some level of local community participation, Gurugram, to some extent, has led to exclusion. In Kaikondrahalli Lake, local people are allowed to harvest grass from the edge of the lake and fishing is allowed on the basis of contracts given out by the fisheries department. What then is the role of communities which may have had traditional rights over the piece of land before the urban restoration work was undertaken? In both cases, it seems that while the co-management models have been taken over by the upper class, it is often the poorer sections whose rights get extinguished. Nagendra compares it to the debate of rights of local communities in a rural setting—'It often becomes like a situation in a protected area, where traditional rights like grazing of cattle by local communities is stopped simply because the state does not allow it.'

Perhaps, the bigger service urban restoration efforts provide is connecting people to nature—which otherwise may never happen in cities. Does this, in any way, make urban dwellers more sensitive to Earth or committed to bigger environmental causes? We don't know. The people interviewed in these case studies seem to say yes. Latika, who spearheaded the creation of the Aravali Biodiversity Park in Gurugram, admitted that she was more careful now of how much water she consumed in her house, that she composted her waste and recycled as much as she could. So these two projects indicate that, at the very least, such urban restoration efforts are making citizens more aware of their role as consumers of natural resources. If the sheer volume of people in cities who followed Latika increased, a lot would change, especially in terms of the way we do business—from our packaging to consumption of plastics or chemicals. The effort would have a radical impact, beyond just a lake or a park that a handful of citizens managed to restore.

That brings us to reviewing urban rewilding in the larger context of rewilding and rehabilitating our forests and its wildlife—and begs the question: Is urban rewilding even worth the effort? Some people have argued that it is better to invest money in protected areas that are home to a higher percentage of wild, endangered species. They believe that investing in urban restoration efforts is a waste of time, as it doesn't help the species that truly need saving. There are others, however, who argue that creating spaces for biodiversity in urban areas is crucial, as it brings people closer to nature. The pigeon paradox rightly sums up this dilemma, as elucidated by Robert Dunn and his co-authors in their paper in *Conservation Biology*. They noted that species and ecosystems won't be saved in cities, but they will be saved by the votes, leadership, and decisions of those living in the city. Therefore, a great deal of future conservation will rely on the interactions of urban city dwellers with nature found in the cities—and these may sometimes include non-natives like feral pigeons.

The pigeon paradox is based on three simple assertions: (1) that the current conservation action is insufficient, (2) people are more likely to take conservation action when they have direct experiences in the natural world, and (3) as human populations (and hence sources of conservation action) shift to cities, humans will primarily experience nature through contact with urban nature. The nature of urban restoration ecology, as elucidated by the pigeon paradox, may focus on species that may not be on the top 10 list of most conservation biologists. But proponents of urban restoration argue that at least an involvement in city-focused restoration would make people sympathetic or malleable when it comes to rallying public opinion around saving ecosystems and species that are in urgent need of that kind of attention.

Fortuitously, both the examples discussed in this chapter escape the pigeon paradox, as they focus on species that had vanished from the urban landscape while, at the same time, creating a lobby for wider conservation goals. There are examples from other parts of India where the pigeon paradox comes into play. This can be seen in the context of community examples of cleaning up of local parks and creating green spaces that may only provide a haven to common species like crows or squirrels, but are urban citizens' only source of interaction with nature.

The urban jungle, with its many non-native species, may well be the breeding ground for future environmental action. And if urban ecological restoration means saving species which may not be endangered, so be it.

9
Rewilding the Blue

The crisis gripping our seas in the form of depleting marine life, the mass bleaching of coral reefs, and the influx of plastic is an environmental apocalypse that has, by and large, escaped conservation rhetoric in India. This, in spite of the fact that the country has a coastline of over 7,000 square kilometres, more than 40 lakh fishermen who live off the sea, and a long maritime history that dates back to the Indus Valley Civilization.

Yet, I did not come across many 'rewilding' projects that involved our marine-protected areas or any conservation project that seeks to restore the damage we have done to our seas. Does this mean that all is well with the species that inhabit the blue world? Marine biologist Divya Karnad, who received the Future for Nature Award in 2019 for her work with endangered sharks, rightly points out that the concept of rewilding, when it comes to the seas, requires a drastic rethink: 'Because ocean life is so dependent on the medium, i.e. water, and its dynamic properties, the concept of restoration has to be completely re-imagined once you are underwater. In the water, the ecosystem builders are sometimes corals or seagrass in shallow waters, and plankton are just about everywhere else in the ocean. Vegetation and mammals, the two groups that we understand well in terms of growth and reproduction, play a very small part. So essentially, a useful restoration project will have to address the creation of healthy phyto- and zooplankton.' And Divya is right. While terrestrial restoration projects have defined boundaries, such as a national park or sanctuary, where the animals are released, the medium of the ocean and its vastness make restoration or rewilding projects a challenge. This explains why there are such few efforts to restore marine biodiversity or their captive breeding.

In my career as an environmental journalist, there were different opportunities to engage with marine issues. In 2010, I got on a fishermen's boat to report on a disastrous oil spill following a collision between two merchant ships off the coast of Mumbai. Even as efforts were on to contain the damage, tar balls were showing up on sandy beaches hundreds of kilometres away from the accident site—proof of just how bad the devastation was. On another assignment, I travelled with conservation biologist Deepak Apte, now Director of the Bombay Natural History Society, to explore the magical microniche between sea and land, documenting the biodiversity of inter-tidal rock pools by diving in them and pulling out creatures of different shades and sizes like a magician pulling out a rabbit from a hat. I observed bright-coloured crabs clinging for their life to the side of the rocks, and sea anemones, periwinkles, and marine algae in the sunlit areas of the rocky pools near Ratnagiri in Maharashtra.

On another assignment, in Rameswaram, Tamil Nadu, I was dazzled by the aquamarine blue of the Gulf of Mannar Marine National Park and saw a beach laden with stingrays being harvested by fishermen. In Kochi, Kerala, at a local jetty, I witnessed thousands of kilograms of fish being hauled out of the sea by nets that were the size of a football field. I drank tea with fishermen along the coast of Gujarat who complained about having to go out for longer hours in their boats because the mechanized trawlers left very little fish behind for them. In a tiny hamlet in Kollam, Kerala, I met a fisherman who recalled a time he would dip his oars into the waters to listen to the 'song of the fish' and then throw his net in for the day's catch. Ancient wisdom gathered and handed down for generations had now been rendered worthless with modern technological advances, he complained.

Most of us are far removed—literally—from the situation in the seas, and perhaps this explains why efforts to restore marine diversity have been sporadic, isolated, and with impacts that are limited by scale. In this chapter, I examine some of these efforts, patchy as they may be—planting seagrass to bring back the sea cow, creating artificial coral reefs, and reintroducing olive ridley turtles in the sea. These efforts are virtually negligible in scale when compared to interventions made to save land-based species, but they are worth analysing, as they red-flag the crisis facing our water world.

◎

COMMON NAME: Seagrass [English]
SCIENTIFIC NAME: *Cymodocea serrulata* and
Syringodium isoetifolium
CONSERVATION STATUS: Least Concern

◎

The Pamban Bridge, off the coast of Tamil Nadu, and India's first sea bridge built more than a hundred years ago, connects the island of Rameswaram to the rest of the country. Here, a unique experiment in rewilding is under way, but to witness it, you have to dive deep into the ocean.

The town of Rameswaram is the gateway to the Gulf of Mannar—one of the richest regions in the world in terms of marine life. The gulf extends from Rameswaram in the north to Kanyakumari in the south and consists of a chain of 21 islands. The region was declared a Marine Biosphere Reserve in 1989 under the Man and Biosphere Programme of UNESCO, and recognized as a marine-protected area by the Indian government. The biosphere reserve comprises estuaries, beaches, seagrasses, coral reefs, salt marshes, and mangroves. Among the gulf's 3,600 animal and plant species are the globally endangered sea cow (*Dugong dugon*) and six mangrove species endemic to peninsular India. Once found easily in India's waters, the dugong has now become rare, and scientists connect its disappearance to the vanishing of its grazing areas.

Dugongs are almost exclusively herbivorous and rely on seagrass and algae for food. They are slow-breeding animals with a long life, and extreme sensitivity to changes in their environment makes them more vulnerable to extinction. Even with the most favourable conditions, their population is likely to grow at a slow rate. As is often the case with species and their habitats, not only do dugongs rely on seagrass meadows, but their grazing helps propagate certain seagrass species. The absence of dugongs could, therefore, have a profound effect on the marine environment and on the habitat necessary for other species to thrive. That is why any attempt to bring back the dugong must look at the restoration and revival of the seagrass ecosystem.

In fact, seagrass communities are vital not just for the dugong but a host of marine species, including coral reefs. A study conducted by a group of marine biologists, including Diraviya Raj in 2010, records as many as 13 species of seagrasses in the Gulf of Mannar. The authors observed: 'Diverse group of animals, such as sea horses, sea turtles, sea cucumbers, sea urchins, star fishes, gastropods, bivalves, ascidians, sponges, crustaceans are abundant in the seagrass meadows, making the Gulf of Mannar a biodiversity rich zone. Authorities managing the marine national park reported a population of more than 100 dugongs in the region based on a survey conducted between 2007 and 2009.

Once covering more than 60,000 hectares in the Gulf of Mannar, the death knell has been sounded for seagrasses because of various marine activities. The damage has not been limited to Tamil Nadu, with researchers saying as much of 35 per cent of seagrass beds in the country have been destroyed in the past three decades.

Keeping in mind the devastation in the region, different scientific groups launched efforts to plant seagrass species on an experimental basis One such effort was by the Suganthi Devadason Marine Research

Institute (SDMI) in Tuticorin, Tamil Nadu, who initiated a pilot project for seagrass rehabilitation activities in 2008. The area of focus was a small patch of degraded area in the Gulf of Mannar and Palk Bay (located to the southern end of the Palk Strait that separates Tamil Nadu from Sri Lanka). In 2013, the same group completed the rehabilitation of one square kilometre of degraded seagrass area in the Tuticorin Coast with support from GIZ, a German development agency based in Bonn, under the supervision of the state's forest department. In 2014, another similar-sized plot in the same area was identified for rehabilitation, this time with the assistance of the IUCN.

These experiments involved diving to the bottom of the sea and planting 400 sprigs of two seagrass species—*Cymodocea serrulata* and *Syringodium isoetifolium*—by burying PVC frames on the seabed and tying the frames with jute ropes. The growth of the transplanted seagrass was monitored periodically. The project received some setbacks when the grass was damaged by fishing nets, but the researchers seemed optimistic that the rehabilitation project would flourish and help convert dead seagrass sites into healthy, thriving beds that would benefit the dugong and other sea life.

Seagrasses are the only angiosperms that successfully grow in tidal and sub-tidal marine environments.

Typically, they grow in areas dominated by soft substrates, such as sand or mud. They require an abundance of light because of their complex below-the-ground structures. As a result, they are normally abundant in shallow waters. Ocean-bottom areas that are devoid of seagrass are vulnerable to intense wave action from currents and storms. The extensive root system in seagrasses, which extends both vertically and horizontally, helps stabilize the sea bottom in a manner similar to the way land grasses prevent soil erosion. With no seagrasses to diminish the force of the currents along the bottom, beaches and buildings close to the coast can be subject to greater damage from storms.

Seagrass meadows also serve as nesting, resting, or grazing areas for a diversity of marine life; some, such as seahorses and lizardfish, can be found in seagrasses throughout the year, while others shelter in them during certain life stages. Creatures like dolphins are often found feeding on organisms that live in areas with seagrass.

The current threats to seagrass are both anthropogenic and natural. The most common threats identified by SDMI scientists during the assessment were disturbance due to fishing, in particular, sailing trawlers, gill-net operations, and boat anchorage. Because of bottom trawling, tonnes of seagrass is swept ashore every day. But there is another factor that is hindering the growth of seagrass and coral reefs—the cultivation of a seaweed known as *Kappaphycus alvarezii*. This exotic red algae species is now on the verge of becoming a bio-invasive, significantly altering the marine ecosystem in the Gulf of Mannar. Apart from its utility as a source of food, food derivatives, vitamins, proteins, etc., this exotic seaweed provides the raw material for many agar and algae-based industries, which explains its rampant cultivation for commercial use.

Fragments of *Kappaphycus* from cultivation sites have started to overwhelm the seagrass, so what was once a seagrass-dominant area has become algae-dominant. In just five years, seaweed cultivation in the Palk Bay has invaded over two square kilometres of coral reef areas in the Mandapam coast off Gulf of Mannar. And the impact is now visible in the form of broken seaweed fragments smothering and killing corals, and blocking out sunlight within six months of the invasion.

Seaweed is grown in three different ways. One of the most popular in India is using the Single Rope Floating Raft (SRFR) method (Coir Rope & Nylon Rope). This involves building tethered rafts and spreading them out in the sea. Seaweed cultivation that uses rafts over the seagrass beds reduces light penetration, essential for seagrass growth and health. The result: stunted growth and lower shoot density in seagrasses, and a turbid environment, and thus less associated biodiversity, particularly in fish. That's why seaweeds are considered destructive for the marine environment.

All these threats were kept in mind when the restoration efforts were initiated. Local fishermen were employed by SDMI for the restoration work: the youth selected for the project were first made aware of the importance of seagrass and then trained to monitor the survival and health of the rehabilitated seagrass. The site was regularly observed, and after nine months of the transplantation, the average survival rate recorded for *Thalassia hemprichii, Cymodocea serrulata*, and *Syringodium isoetifolium*, the three species planted, was 81.5 per cent, 85.7 per cent, and 78.6 per cent respectively. Perhaps, the best indicator of success was that the micro-organisms that live close to the ocean bed increased in density with the increased seagrass cover. However, consultations with the local fishing community at large to reduce the fishing pressure in and around the site could lead to even better results, and also contribute to more long-term restoration efforts. In a welcome move, the forest department began to uproot *Kappaphycus* from the reef area of the Gulf of Mannar Marine National Park to reduce further impact and stress, and to supplement the work done on planting seagrass.

Though the effort to rewild the seas off the coast of Tamil Nadu has seen positive results, it is a long way off from significant conservation success. In order to restore seagrass meadows, or increase the presence of indicator species like the dugong, a comprehensive programme that involves working with the local fishing communities is needed, coupled with the removal of large tracts of invasive seaweeds. In 2018, the Gulf of Mannar Marine National Park authorities sent a proposal to the ministry of Environment, Forest and Climate Change, seeking funds for seagrass rehabilitation. The project aims at creating habitats and source of food in shallow water for dugong. If the dugong is to be saved, seagrass meadows need to be restored.

COMMON NAME: Finger Coral
SCIENTIFIC NAME: *Acropora humilis, Vidruma in hindi*
CONSERVATION STATUS: Near Threatened

The next example of rewilding is from the coast of Gujarat. It provides important lessons for conservation interventions, most importantly, that a 'Plan B' is crucial for any restoration effort.

In the industrial township of Mithapur, a private chemical company decided to partner with an NGO to revive dying coral reefs in 2008. A habitat survey of the Mithapur reef in the early phase of the project found less than 12 per cent of live hard corals, highlighting the need to focus on their recovery and restoration. Approximately 0.56 square kilometres of the Mithapur reef falls under Gujarat's Marine National Park (MNP). The *Acropora humilis* coral, once part of the coral reefs off the Gulf of Kutch, is now locally extinct, and only dead skeletons can now be seen. The Gujarat forest department and WTI, through a partnership with Tata Chemicals, which has a presence in the area, transplanted fragments of locally extinct corals in the coastal waters of Gujarat. Fragments of the *Acropora* species were collected from as far as Agatti Island in Lakshadweep, where they are common, and moved about 1,500 kilometres for reintroduction in the Mithapur reef in the Arabian Sea and Poshitra Reef of the MNP in the Gulf of Kutch.

Coral reefs are highly productive marine habitats. They are indicators of the quality of marine environments and are of great significance for the livelihoods of coastal communities: coral reefs support high diversity of fish life, which in turn helps fishermen.

The aim of the Mithapur Coral Reef Recovery Project was to strengthen this marine ecosystem. The intention was to map reef boundaries, raise a coral nursery, set up protocols for their long-distance transportation, and transplant them in the sparsely populated reef areas, as in the waters off Mithapur.

In order to understand the restoration process, it is important to know more about corals. What we usually understand as a coral is actually a group of many individual, yet genetically identical, multicellular organisms known as polyps. Some corals can catch small fish and plankton using stinging cells on their tentacles, like those in sea anemone and jellyfish. Most corals obtain the majority of their energy and nutrients from photosynthetic unicellular algae that live within their tissue. The polyp's tentacles immobilize or kill prey, then contract to put it in the stomach. Once the prey is digested, the stomach reopens, allowing the elimination of waste products and the beginning of the next hunting cycle.

There are two kinds of corals: hard and soft. Hard corals (Scleractinia), such as brain, star, stag horn, elkhorn, and pillar corals, secrete rigid calcium carbonate exoskeletons that protect their soft, delicate bodies. Soft corals, such as sea fans, sea whips, and sea rods, lack exoskeletons.

Coral reefs are extensive networks of calcium carbonate fused together through adhesives secreted by the polyps, resulting in the formation of thin plates and layers over time. The average lifespan of a polyp in the wild is two to hundreds of years and that of a colony is five years to several centuries. Their protection status is currently endangered. In India, four major coral-reef ecosystems exist, in the Andaman & Nicobar islands, Lakshadweep islands, the Gulf of Mannar off Tamil Nadu, and the Gulf of Kutch off Gujarat. The Gulf of Kutch is, in fact, one of the rare places in the world where you can look at corals without having to dive underwater.

Based on their shape, corals can also be classified as branching, boulder, and foliose. Ecologically, all three forms are important; however, branching corals are considered more so. This is because even though they aren't very sturdy, they grow faster and can form an intricate network of spaces where fish can thrive. Boulder and foliose species, on the other hand, are slow growers but very sturdy.

For a complex set of reasons, the branching corals off the Gulf of Kutch went locally extinct, and WTI has been attempting to bring them back. Explains WTI marine biologist Sajan John in an email interview, 'The experiment involved first studying the existing threats. It was observed that during the four months when there is a ban on fishing, local fishermen engaged in "poison fishing" on the reef. They pour an extract of a cactus (Euphorbiacae) sap in the water to poison fish and sell them in the market. This is a major threat to the survival of coral reefs; it reduces coral reef biodiversity, as [the sap] affects not just the target fish but other marine organisms too.'

The experiment to transplant the corals in the Mithapur reef could be the longest-distance transportation, translocation, and transplanting attempt of corals in the world. Fragments from an identified donor *Acropora* colony in Lakshadweep were broken and transported to the recipient site by following a scientific protocol. Such transported fragments, it was hoped, would propagate into a new colony at the recipient site, so as to reproduce and repopulate the recipient reef. Factoring in the distance and complex logistics of transportation, and after a thorough survey of scientific literature, the translocation of the species from Lakshadweep was considered the best option for the reintroduction experiment. However, the effort was not successful. In the first attempt, the corals could not survive the heavy sedimentation at the new location, and all of them died. The second attempt, too, was a failure—all the coral fragments died en route.

There were, however, some significant learnings. It allowed the researchers to take stock of the status of corals in the area and their health. The biological surveys done in the course of the project revealed that the Mithapur region still had some important life forms.

The WTI team was able to document 29 coral species, 55 species of fish, 28 species of seaweed, 13 species of crabs and lobsters, 26 species of seashells, nine species of sea slugs, five species of flat worms, and 12 species of other associated animals. Reef-habitat mapping helped identify key degraded areas. It also allowed for the identification of the best area within the reef that could harbour nurseries for coral transplant. Further, one sea slug and one seahorse species were also found for the first time in India.

After the two failed attempts from 2008 to 2013, the translocation of the *Acropora* species was halted, and WTI decided to come up with a Plan B. Instead of getting corals in from Lakshadweep, the focus shifted to restoring the area with local species of boulder corals, and about 400 artificial reefs and 53 coral garden nurseries were created. The artificial reefs (made of limestone boulders) provide a conducive habitat for the natural attachment of polyps and also protect the coral garden nurseries within. These artificial reefs also function as a 'fish house', harbouring reef as well as coastal fish species, thus increasing fish production.

A similar initiative was undertaken by the forest department of Gujarat along with the Zoological Survey of India (ZSI) in 2015, this time inside the Marine National Park, transplanting corals brought in from the Gulf of Mannar on the Southeastern coast of India. Explains John, 'The reason that WTI did not want the corals from Gulf of Mannar was [because] the Mithapur project is along the west coast of the country. We wanted species from the Arabian Sea. That is why we took coral fragments from Lakshdweep.' The ZSI project, on the other hand, used fragments from the east coast and brought it to west coast of India.

In the ZSI project, 215 artificial cement triangles were transplanted with 1,569 coral fragments in a small tide drain channel in the Pirotan Island, Marine National Park, and the Gulf of Kutch. The overall survival rate of the transplanted corals was 77.57 per cent. According to the ZSI, coral fragments even overcame the impairment of the water current, sedimentation, algae cover, and wave actions.

No other effort has been undertaken to rewild the waters with corals. What is needed is a mass effort to increase the scale of operation if coral reefs are to be restored to their former glory. The problem with such experiments is they take away from the real issues that caused the death of the corals in the first place. While climate change and rising sea temperatures may be too complex factors to address at the local level, others are not. These include the discharge from polluting industries in the vicinity and better regulation of the activities of the fishing industry.

But there are advantages to small-scale initiatives as well, says John: 'The project created awareness within the communities on the importance of corals and the destruction that bleaching powder can cause on reef and reef life. The artificial reefs helped in attracting more fish to the area, which, in turn, helped fishermen nearly double their catch. This helped the NGO to win local support.'

In Tamil Nadu, coral rehabilitation was first initiated in the Tuticorin coast of Gulf of Mannar in 2002 by SDMRI with the support from ministry of Environment and Forests and Coral Reef Degradation in Indian Ocean (CORDIO). The researchers from SDMRI successfully standardized and field-tested the low-cost and low-tech transplantation technique using artificial substrates like cement frames and multipurpose fish houses with native coral species. In 2013, SDMRI completed rehabilitation of one square kilometre degraded reef area near Vaan Island in the Tuticorin coast with support from GIZ under the supervision of the Department of Environment, government of Tamil Nadu. The overall survival of the coral in the region was over 80 per cent, and fish abundance increased from 34 per cent to 65 per cent in five years.

Says marine biologist Divya Karnad, 'The Gujarat example is the only one in India where live corals have been transplanted. Its low rate of success and high cost ensured that this method was not repeated elsewhere. The approach to set up artificial reefs along the coast of Gujarat and Tamil Nadu has met with more success. They have generally been successful at attracting fish from surrounding areas, but a few have also begun to act as substrate for coral. It may take a few more years before we can start seeing proper colonization and breeding at these sites.' But none of it will be enough if we don't think about why the reefs needed rehabilitation in the first place, and take measures to course correct.

◎

COMMON NAME: Olive ridley turtle [English], kouchop koutohul [Odiya]
SCIENTIFIC NAME: *Lepidochelys olivacea*
CONSERVATION STATUS: Least Concern

◎

While the previous two examples concerned habitat restoration at the bottom of the sea, let's look at efforts to rewild a marine species. In 2006, I was invited by Greenpeace India to report on the threats that olive ridley turtles were facing along the coast of Odisha, where they come en masse for nesting. While the images of hundreds of turtles coming from the sea to lay their eggs are iconic, tragically, I witnessed only dead turtles. Every beach we walked to was strewn with carcasses of olive ridleys, as the crashing waves nudged the dead animals further landwards. That year was not an aberration; dead turtles had become a common sight every year during the mass nesting season in Odisha.

The beaches where these gentle giants of the sea come to give birth had become virtual death traps for them. A rough estimate by Greenpeace India put the figures at more than 1,00,000 dead turtles being washed ashore in just a decade. The deaths were attributed to a range of factors, the predominant one being uncontrolled, mechanized fishing in areas of high sea turtle concentration. As the turtles would approach shallow waters to nest, they would get entangled in the fishing nets; others ended up drowning, unable to free themselves from the massive nets, leading to high mortalities.

The olive ridley is the most widely occurring species of sea turtles that nest at several sites in the western Indian Ocean, the Indian subcontinent, and South East Asia. Though the olive ridleys nest in many places along India's coast, their mass nesting takes place only in Odisha. This species is well known for its synchronous nesting behaviour—and thus is known as *arribada*, Spanish for 'arrival'.

The 480-kilometre coastline of Odisha harbours three mass nesting grounds—the Gahirmatha rookery along the northern coast, the Devi rookery 100 kilometres south of Gahirmatha, and the Rushikulya rookery near the mouth of the Rushikulya River along the southern coast. Between January and May every year, more than half a million olive ridley turtles descend here. All these places are densely populated by fishing communities, and intensive fishing takes place in the offshore waters of these rookeries during breeding season. In Hindu mythology, sea turtles are worshipped as an incarnation of Lord Vishnu, and hence most fishing communities along the coast do not consume turtle meat or eggs, or harm turtles in any way. But it is the entanglement with the fishermen's nets that creates the problem. Kartik Shankar, a wildlife scientist who has worked extensively in the region on olive ridleys, observes in his book *From Soup to Superstar* that 'using a single species as a flagship of conservation can be a double-edged sword. On one hand, it can bring in support and protection measures but on the other hand can alienate local people as it impacts livelihoods or restricts access to traditionally used areas'. That's why it was only appropriate that any intervention to save the turtles involved local fishermen.

It was the research work by a scientist that motivated Rabindranath Sahu, a local fisherman in Odisha, to start a people's movement for the turtles. In 1994, Bivash Pandav visited the village of Purunabandha on the banks of the Rushikulya. The wildlife scientist was appalled to see turtle eggs destroyed by crows, dogs, and jackals. A chance walk with Pandav had made him a turtle champion for life, when he observed more than 30,000 turtles lay their eggs in a single night. The image stayed with him forever and he decided to set up a community group of turtle protectors.

The lack of awareness about olive ridley turtles and their mating, nesting, and hatching habits among the villagers prompted him to establish the Rushikulya Sea Turtle Protection Committee (RSTPC) in 1998 with the assistance of local fishermen. The RSTPC started off with activities that included keeping the beach and nesting area clean, monitoring the arrival of turtles, and keeping fishermen and predators away from the beach during the nesting and hatching season. Gradually, the activities of the RSTPC grew to include awareness programmes for the local population, including school children. At the Rushikulya beach, baby turtles would often get disoriented due to artificial lighting and find themselves getting lost or heading towards land instead of the sea. The RSTPC members would then collect the lost hatchlings in buckets and release them in the ocean.

Due to the efforts of Sahu's organization, the turtle is better protected in the villages of Purunabandha, Gokharkuda, Podampeta, and Kantiagada. The members, who have been trained by wildlife experts, serve as research assistants, working with turtle scientists like Bivash. The hard work put in by the wildlife scientists and the local youth they have involved in the project has ensured that the village of Podampeta has acquired global status as a 'turtle village'. For more than a decade now, Sahu and his group of volunteers have cleared away old nets and debris before the turtles reach the sands, remind fishermen to stay away, and scan the beach for the first sporadic nesters.

Happily, fishing communities from other parts of the country have also become sensitive to the plight of turtles at large. In 2002, a small fishing community in Maharashtra started active protection work for olive ridleys in Velas, a tiny village on the northernmost boundary of the district of Ratnagiri. In the first year, Sahyadri Nisarga Mitra (SNM) undertook protection work in one village and successfully protected 50 nests. Within a short span of time, SNM spread the protection work to the entire coast of Maharashtra, which amounts to about 720 kilometres of coastline.

In the next four years of the project, they arranged protection work in 15 villages. Even with their limited resources, the fishermen had succeeded in returning to the sea as many as 7,610 hatchlings within four years.

In terms of the scale of the problem, the efforts by the communities in Velas and Rushikulya to return the hatchlings to the sea may seem like a drop in the ocean, but this has had collateral advantages. They have succeeded in changing the image of the fishing community, from being destroyers to that of protectors. It is now the fishermen who protect the nests from by feral dogs or human activity, and have become the champions of olive ridleys. This, in turn, has prompted conservation efforts not just for turtles but other species as well. In fact, in Maharashtra, SNM has turned protection efforts to other species in the area, such as the sea eagle and the Indian swiftlet. Furthermore, conservation has dovetailed with tourism, with homestays created by SNM for tourists during the turtle-hatching season.

The strength of these examples is that they typify the first marine rewilding effort that put the fishermen at the centre of conservation interventions. The physical proximity of the fishermen to the nesting beaches and the partnership with scientists created the right ingredients for success. The rewilding effort by fishermen is a gentle nudge in the right direction for protecting marine life.

◎

There are no reintroduction programmes that involve captive breeding of marine species as such in India. The examples outlined in this chapter involve the restoration of a marine habitat in the first two cases, and in the last one, that of a species. However, all the projects were limited by scale and impact, especially when compared to efforts in restoration ecology involving land-based species or habitats. Nonetheless, they indicate a start in the right direction. Biologists like Deepak Apte suggest that captive breeding and rewilding efforts are needed for species like the giant clams in the Andaman & Nicobar islands, where they have declined by over 90 per cent and face local extinction on many island reefs. There are other projects as well, says Apte, focused on the revival of species like the dugong and the Gangatic river dolphin, which are currently under consideration with the government.

Karnad rightly concludes: 'Unlike the clearly demarcated patches of land that one can restore, ocean restoration is necessarily dynamic in space.' That, perhaps, explains why rewilding projects of the sea are limited in their scope and scale.

Of course, it must be kept in mind that the crisis facing our seas is not confined to India alone. In 1988, a giant soupy accumulation of trash was found floating in the sea between Hawaii and California that was subsequently labelled as the 'Great Pacific Garbage Patch'. Estimated to be three times the size of France, the floating mess is a visual reminder of the need for a healing touch—and for a greater application of the theory of rewilding to our blue world.

10
Restoring the Ark to Park Continuum

This book restored my faith in conservation. In the decade of environment journalism that I had behind me, I had become too used to focusing on the bad news—the forest that was destroyed to give way to yet another expressway, the poachers who got away, the elephant calf whose mother had been mowed down by a speeding train, the communities that lost their homes to a mining project. A country that is home to megadiversity, yet seeks to fell its forests to megadevelopment projects can leave a tree hugger frustrated and disappointed.

In the process of travelling the country documenting stories of rewilding, I was able to admire the grandeur of the one-horned rhinoceros, gaze with wonder at the eggs of the mahseer, walk in the footsteps of river turtles on hot riverine sands, and then feel the cool waters on my toes as we released the hatchlings in the Chambal. Following these rewilding projects proved a balm to my soul—and I had to constantly remind myself that this bold new approach to conservation has its share of limitations.

Rewilding is certainly an ambitious alternative to current approaches to conservation, and interest in it is growing in popular and scientific literatures. India may not have embraced the term formally, the way Soulé and his colleagues envisaged it, but there is no doubt that as a tool for conservation it is being used across the country. That is obvious from the examples outlined in this book. From the national parks of Assam to the dying lakes of Bengaluru, from species as diverse as the one-horned rhino to coral reefs at the bottom of the sea, forest officers and conservation practitioners are accepting and using the term, adapting it to their way of working. For every case study described in this book, there were many more—though they were small, isolated, they were outstanding projects. But they were either too limited by scale, or seemed to be focused on the rescue of individual animals rather than saving a species as a whole. A prime example is the effort by an animal-welfare organization to release an injured deer back in the wild or a couple from the city who bought a piece of land to plant trees on. The interest for this book were conservation efforts that focus typically around saving species or restoring ecosystem processes to scale.

One common thread in all the case studies discussed in this book was the engagement (or lack of it) as far as local communities were concerned. Rewilding efforts, laudable as they maybe, do raise a series of ethical concerns, from the lack of consultation with local communities to their conflict with more established forms of environmental management.

The absence of emphasis on the co-option of local communities has been the missing link in most conservation projects in India. While this trend is now changing, especially among the younger generation of wildlife scientists, the fact is that most of the projects surveyed in this book tended to ignore local communities; even if they were included, they were not central to the decision-making process. Given that, in our country, biodiversity survives in a sea of humanity, this may be a huge oversight. Most projects choose the easy way out of releasing species in protected areas, assuming they are already free of human pressure. In fact, a rewilding effort that involved releasing crocodiles in the Neyyar River in 1983 failed precisely because of the failure to consult with the local communities. Thirty-six mugger crocodiles were reintroduced into the reservoir of the Neyyar Wildlife Sanctuary in 1983 as part of the crocodile conservation project launched in Kerala with the government of India, the United Nations Development Fund, and the Food and Agriculture Organization of the United Nations. Crocodiles, though present, were rare in the Neyyar river system before the reintroduction programme, but the future of the population was bleak because of the animosity of the local population. From 1985 onwards, the crocodiles started attacking local inhabitants along the bank of the reservoir, so further reintroductions of the mugger were stopped. Scientists who subsequently reviewed the initiative 18 years after the reintroduction effort estimated that there should have been more crocodiles in the river system, however nine had to be removed due to conflict with people. They further recorded that 29 incidents of crocodile attacks on humans were reported prior to the study and six occurred later, including two fatalities. As local people utilized the reservoir for various purposes, they did not support the conservation of crocodiles there. The case study indicated the failure of the reintroduction programme of the muggers in the Neyyar reservoir.

Fortuitously, some of the case studies discussed in this book—for instance, the one on rewilding vultures—escape this criticism. The efforts to bring back vultures to our skies have kept agrarian communities as the central focus of conservation interventions. Creating VSZs required the farming community to stop using diclofenac, the lethal drug that wiped out the birds. So the project stakeholders had to work closely with them to develop a comprehensive communication plan and get them to start using alternatives. Others met limited success for lack of stronger participation from the local communities, such as the project of planting seagrass at the bottom of the seabed. Then there are those involving species like the rhinoceros in Manas that have attempted to enlist community participation by handing out doles like gas connections, but failed to elicit any real co-operation—and possibly concern—for their conservation goals.

There are other problems with rewilding as it is being applied in India today. Of the 'three Cs' that Soulé states as essentials of rewilding, the recognition of corridors has been the weakest, as they have no legal protection. The chapter on the tigers of Panna elucidates a lack of policy on tigers that use these corridors to move out of the national park. The need to dart the tiger that 'strays' out of the park and bring it back for fear that it may get into conflict with humans is questionable. Says tiger expert Raghu Chundawat on corridors: 'There is a reason why we don't have fences for our protected areas; we want

dispersal because we know connectivity is important.'
And then there are other problems with corridors. Most
parks in India exist as isolated islands of biodiversity
and the migratory routes that once functioned as
opportunities for wild species to move out have today
become expressways to speeding traffic or given way to
craters left behind by years of coal mining. The result:
corridors are now virtual death traps for the leopard,
tiger, or elephant that may be using them, thus closing
down any opportunity for wild animals to move to other
habitats. This may be the reason why the Roadkills app
was developed by the Wildlife Conservation Trust in
2018; it enables citizens to share photos and locations
of animals killed on highways across India. 'Unplanned
development of roads and railway lines is the major
cause of wildlife roadkills,' says wildlife biologist Milind
Pariwakam and points that as the prime motivation for
developing this app. Within just four months of having
been active, the app was used by more than 3,000 people
who shared more than 900 instances of wild animals and
birds being killed on highways.

According to Pariwakam, an analysis of the data on
roadkills reveals mammal species like the tiger, leopard,
hyena, wolf, jackal, fox, jungle cat, fishing cat, wild pig,
hare, grey mongoose, langur, rhesus macaques, small
Indian civet, palm civet, spotted deer, nilgai, sambhar,

blackbuck, rodents, bats, and palm squirrels; as many as 26 species of reptiles that include the monitor lizard, cobra, Russell's viper, saw-scaled viper, common krait, Indian rock python, oriental wolf snake, sand boa, rat snake, barred wolf snake, and cat snake; and over 30 species of birds such as the tailorbird, small green bee-eater, nightjar, shrikes, red-vented lapwing, jungle owlet, vulture, griffon vulture, egret, magpie robin, babbler, parakeet, and buzzard. Clearly, roads that cut through wildlife areas and corridors have a deleterious impact, and this citizens' initiative is helping document its scale.

Morbid as the task may be, the information generated from the application can help identify crucial sections of roads or railway lines where animal deaths are high, and so pinpoint regions that require urgent attention. The data can also help determine what species is more at risk on specific road or rail stretches, and plan the ideal measures for the location—from underpasses or overpasses for large mammals to canopy bridges for arboreal ones such as monkeys. The future success of rewilding will be incumbent on keeping corridors safe to ensure genetic viability.

In fact, that is exactly what biologists Sandeep Sharma et al. at the Smithsonian Conservation Biology Institute noted in a 2013 paper in the *Journal of Ecology and Evolution*. Using felid faeces to gauge the genetic diversity of tigers and leopards in central India, the authors found that natural forest corridors between the parks were essential for genetic exchange and vital for the future of the species. Of 1,411 fecal samples collected in the protected areas and along the pathways between them, the researchers were able to assign 463 to tigers and 287 to leopards. The samples recorded the background of each cat, giving the zoologists an outline of how much genetic diversity exists in the populations and whether cats from different forest patches were mating with each other. The tiger study showed that they had not gone through a genetic bottleneck that would be expected for a population confined to spread-out conservation areas. In fact, they showed a high amount of genetic diversity. The authors attribute this to the wildlife corridors that allow tigers to travel past mining operations and roads in search of their neighbours. For central India's tigers to survive, the researchers argued, the pathways that allow the felids to travel and inter-breed must remain open. Habitat fragmentation can divide populations of species into isolated groups, which can lead to in-breeding and a genetic bottleneck, affecting the long-term viability of the population. The study thus highlights that creating large patches of forest, which is the traditional approach to conservation, is not enough; *connecting* these habitats is important, so that species can 'meet and greet' and maintain their genetic diversity at the same time.

Even as rewilding has proved a successful conservation tool by organizations and park managers across India, there are species in distress that could benefit from its application. For instance, some argue that the great Indian bustard (*Ardeotis nigriceps*) could benefit immensely from an intervention designed around the theory of rewilding. Once widely spotted across 11 states, the bird is currently listed as critically endangered. There are less than 150 bustards left in the country and each individual death brings the species closer to extinction. Could a captive-breeding project resuscitate this bird species?

However, scientists from the University of East Anglia disagreed. In a paper published in 2015 in the *Journal of Applied Ecology*, they argued that bustards are particularly difficult to keep and breed in captivity and, therefore, merely breeding them in captivity may not help. Described by the scientists as a 'challenging stress-and-injury prone species', large bustards are particularly susceptible to accidents and fractures in captivity, as well as having delayed reproductive maturity and low fecundity. That's why captive breeding alone may not be the answer. But a rewilding programme that could have looked at restoring the grasslands where the bustard is found could well help boost their numbers. In fact, the pygmy hog and the rhino rewilding efforts in the Northeast have done exactly that—restored grassland habitats to enable the species to rebound. A modelling exercise by the same group of scientists predicted that just by implementing in-situ conservation over the next decade would recruit more adult females to the wild within 30 years than a project focused solely on captive breeding.

Apart from fish species like the mahseer, there are a number of other, lesser-known fish that could benefit from rewilding. Ornamental species such as the red-line torpedo barb, popularly known as 'Miss Kerala'—which has a distinct pattern of colour—has spawned a trade for the domestic and international aquarium industry. It is estimated that about 1,00,000 fish are flown outside India each year, making up 65 per cent of India's export of ornamental fish. Notes ichthyologist Rajeev Raghavan, 'Species are selected for captive breeding based on the ease with which they can be bred. The conservation angle is generally missing.' A rewilding project that looked at protecting the rivers of the Western Ghats in southern India where the species is found could be far more useful than simply breeding it in captivity for trade.

Deepak Apte of BNHS points out that marine species like the giant clams in the Lakshadweep and Andamans could also benefit from a rewilding effort. Giant clams, belonging to the family Tridacninae, are the largest living bivalves, and can grow over a metre in length. They have a narrow range of geographical distribution and occur exclusively within the tropical reefs of the Indo-Pacific region. These clams are biologically important; they filter large amounts of water with harmful waste nutrients like ammonia and expel clean water to the environment. But their numbers have crashed from their unregulated harvesting for food and souvenirs, and from habitat decline.

While rewilding, therefore, has been applied to typical charismatic megafauna like tigers and rhinos, some of the lesser-known forgotten species that don't typically receive much conservation attention could also benefit from it. That's the strength of the rewilding theory; it generates conservation optimism while giving humans a mandate to fix the problems generated by them.

Which finally brings us to the last question that this book set out to answer: Is rewilding a Western import?

The birth of the rewilding theory has been, of course, in the West. A cursory glance at some of the biggest projects on rewilding reveal a slant towards European countries or the US, which typically lost most of their megafaunal species or have large tracts of land available for rewilding. As biologist Meredith Bernstein from the University of Oxford observes: 'Interestingly, there has been almost no discussion to date of what rewilding would look like in the global South. We tend to think that the South still has plenty of primary forest and pristine wilderness, constituting conservation priorities: why "re-wild" when it is already "genuinely" wild?'

The fact is that even typical conservation programmes in the global 'south may not be working well because of poaching, habitat fragmentation, or simply the pressures of growing economies. That's why rewilding must not be seen as a prerogative of the developed world—other parts of the world that are rich in biodiversity, that are facing all kinds of threats can also benefit from it'.

The examples discussed in this book illustrate that rewilding is not necessarily specific to the west but what conservation practitioners have done is to adapt and modify it to make it more relevant for us. India may not have large swathes of land—for instance, the 1 million hectares of land committed by European countries for rewilding—or the massive investment of funds to achieve this target. India has remoulded and reshaped the theory. As Jamie Lorimer notes, 'The term rewilding does not have a single simple definition. Instead, it has proved useful as a way of describing an approach to conservation that seeks to maintain or even increase biodiversity and reduce or reverse past and present human impacts by restoring more functional ecosystems.'

Lastly, rewilding cannot be seen as the magic solution to all conservation issues. The experimental nature of rewilding means it comes with no guarantees, and the future natural landscapes created may come with their own associated risks. A number of risks that have been identified with rewilding are about species introduction or reintroduction; for example, depletion of the donor populations, risks of bringing in disease, or low genetic variability among the introduced individuals. Opposition to rewilding is particularly likely where projects are perceived as being imposed from the 'outside', with little consideration for local interests, as happened in the case of the Neyyar crocodiles. The flagship taxa of rewilding tend to be mega herbivores and carnivores, species that generate considerable public appeal and revenues for conservation. A focus on these animals can result in the neglect of other species that may need as much attention.

Perhaps, the biggest lesson for me while writing this book was that Indian wildlife conservation has come of age. It has joined the dots, recognizing that the 'ark to park' continuum is a complex process. Rewilding is no longer about breeding animals in captivity and then setting them free. A consolidated plan to secure habitats and corridors is necessary. As for me, at a personal level, rewilding helped me put the soul back in conservation.

Select References

Aghor, Ashwin, M. Suchitra, Anupam Chakravartty, and Kumar Sambhav Shrivastava. 2015. 'Sand Slips', *DownToEarth Magazine*, June. Available at www.downtoearth.org.in/coverage/sand-slips-37957, accessed on 1 February 2019.

Agrawal, Paankhi, and Mayank Sinha. 2012. *A Report on the Status of Pardhis in Mumbai City*. Mumbai: Tata Institute of Social Sciences. Available at www.castemumbai.tiss.edu/wp-content/uploads/2015/04/Pardhis-In-Mumbai.pdf, accessed on 12 February 2019.

Allen, Irma. 2016. 'The Trouble with Rewilding', *Entitle Blog—A Collaborative Writing Project on Political Ecology*, 14 December, available at www.entitleblog.org/2016/12/14/the-trouble-with-rewilding/, accessed on 1 February 2019.

Barua, Maan, and Tarsh Thekaekara. 'The Politics of Rewilding/Reintroductions: The Lion in India.' Slides, Megafauna and Ecosystem Function: From the Pleistocene to the Anthropocene, Oxford University,

Oxford, March 2014. Available at www.eci.ox.ac.uk/events/2014/megafauna-conference/barua.pdf, accessed on 1 February 2019.

Bernstein, Meredith Root. 2014. 'When Rewilding Isn't Mad: Guanacos can Transform the Espinal of Chile', *The Hindu*, 14 March. Available at www.thehindu.com/sci-tech/when-rewilding-isnt-mad-guanacos-can-transform-the-espinal-of-chile/article5815215.ece, accessed on 1 February 2019.

Cunningham, A.A., V. Prakash, D. Pain, G.R. Ghalsasi, G.A.H. Wells, G.N. Kolte, P. Nighot, M.S. Goudar, S. Kshirsagar, and A. Rahmani. 2003. 'Indian Vultures: Victims of an Infectious Disease Epidemic?' *Animal Conservation*, 6: 189–97.

Cuthbert, R., M.A. Taggart, V. Prakash, M. Saini, D. Swarup, et al. 2010. 'Effectiveness of Action in India to Reduce Exposure of Gyps Vultures to the Toxic Veterinary Drug Diclofenac. 2011. *PLoS ONE*, 6(5): e19069. DOI:10.1371/journal.pone.0019069.

Dolman, Paul, N.J. Collar, Keith M. Scotland, and Robert Burnside. 2015. 'Ark or Park: The Need to Predict Relative Effectiveness of Ex Situ and In Situ Conservation Before Attempting Captive Breeding', *Journal of Applied Ecology*, 52: 841–50.

Dunn R.R., M.C. Gavin, M.C. Sanchez, J.N. Solomon. 2006. 'The Pigeon Paradox: Dependence of Global Conservation on Urban Nature', *Conservation Biology*, December, 20(6): 1814–6.

Select References

Dutt, Bahar. 2018. 'In Modi's Constituency, a Wildlife Sanctuary is Quietly Being Erased', *The Wire*, 24 September. Available at www.thewire.in/environment/in-modis-constituency-a-wildlife-sanctuary-is-quietly-being-erased, accessed on 1 February 2019.

Gee, E.P. 1964. *Wildlife of India*. London: Collins Ltd.

Hunt, G. 2013. 'In Murmurations, Starlings Have a Darwinian Dance Partner', The Cornell Lab of Ornithology *All About Birds* blog, 15 January. Available at www.allaboutbirds.org/in-murmurations-starlings-have-a-darwinian-dance-partner/, accessed on 2 February 2019.

Jayson, Eluvathingal, Sivaperuman Chandrakasan, and P. Padmanabhan. 2006. 'Review of the Reintroduction Programme of the Mugger Crocodile *Crocodylus palustris* in Neyyar Reservoir, India. *Herpetological Journal*, 16(1): 69–76.

Johnsingh, A.J.T., and M. Madhusudan. 2009. 'Tiger Reintroduction in India: Conservation Tool or Costly Dream', in Matt W. Hayward, Michael J. Somers (eds), *Reintroduction of Top-Order Predators*, pp.146–63, Wiley, DOI:10.1002/9781444312034.ch7.

Kumar, J.S.Y, C.H. Satyanarayana, Krishnamoorthy Venkataraman, Imtiyaz Beleem, G. Arun, Chandran Retnaraj, Ram Kumaran, and R.D. Kamboj. 2017. 'Coral Reefs Transplantation and Restoration Experience in Pirotan Island, Marine National Park, Gulf of Kachchh, India', *Indian Journal of Geo-Marine Sciences*, 46(2): 299–303.

214

Select References

Kuppusamy, Sivakumar. 2013. 'Status and Conservation of *Dugong dugon* in India: Strategies for Species Recovery', in K. Venkataraman, C. Sivaperuman, and C. Raghunathan (eds.), *Ecology and Conservation of Tropical Marine Faunal Communities*. DOI: 10.1007/978-3-642-38200-0_27.

Lang, J.W., and P. Kumar. 'Chambal Gharial Ecology Project—2016 Update'. World Crocodile Conference, Proceedings of the 24th Working Meeting of the IUCN_SSC Specialist Group, pp. 136–48. IUCN: Gland, Switzerland.

Lang, J.W., and S. Whitaker. 2010. 'Application of Telemetry Techniques in Crocodilian Research: Gharial (Gavialis gangeticus) Spatial Ecology in the Chambal River, India', in K. Sivakumar and B. Habib (eds), *Telemetry in Wildlife Science*, ENVIS Bulletin, Wildlife and Protected Areas, 13(1): 161–71.

Lorimer, Jamie, Christopher Sandom, Paul Jepson, Chris Doughty, Maan Barua, and Keith J. Kirby. 2015. 'Rewilding: Science, Practice, and Politics', *Annual Review of Environment and Resources*, 40(1): 39–62.

Mallinson, J.J.C. 1971. 'The Pigmy Hog, *Sus Salvanius* (Hodgson), in Northern Assam', *Journal of the Bombay Natural History Society*, 68(2): 424–34.

Menon, Vivek, Rahul Kaul, Ritwick Dutta, N.V.K. Ashraf, and Prabal Sarkar (eds). 2008. *Bringing Back Manas—Conserving the Forest and Wildlife of the Bodoland Territorial Council*. New Delhi: Wildlife Trust of India.

Mukherjee A., H.T. Galligan, V. Prakash, K. Paudel, U. Khan, S. Prakash, S. Ranade, K. Shastri, R. Dave, P. Donald, and C. Bowden. 2014. 'Vulture Safe Zones to Save *Gyps* Vultures in South Asia', *MISTNET*, July–September, 15(3).

Nagendra, H. 'A Tale of Two Lakes: Collective Action in Cities', *SmartCitiesDive* blog. Available at www.smartcitiesdive.com/ex/sustainablecitiescollective/tale-two-lakes-collective-action-cities/71576/, accessed on 27 February 2019.

Nagendra, H. 2016. *Restoration of the Kaikondrahalli Lake in Bangalore: Forging a New Urban Commons*. Pune: Kalpavriksh. Available at www.vikalpsangam.org/static/media/uploads/Resources/kaikondrahalli_lake_casestudy_harini.pdf, accessed on 3 February 2019.

Narayan, G., and P.J. Deka. 2012. 'New Milestone Reached in Efforts to Save the Pygmy Hog, One of the World's Most Endangered Mammals', *Suiform Soundings*, 11(2): 34–7.

Narayan, G., P.J. Deka, W.L.R Oliver, and J.E. Fa. 2010. 'Conservation Breeding and Reintroduction of the Pygmy Hog in NW Assam, India', in P.S. Soorae (ed.), *Global Re-introduction Perspectives: 2010*. Abu Dhabi: IUCN/SSC Reintroduction Specialist Group.

Post, Gerald. 'Evaluation of Tiger Conservation in India: The Use of Comparative Effectiveness Research'. Master's project, Duke University, 2010.

Available at https://pdfs.semanticscholar.org/e1f8/
e6593fb3153b8483046e0f71deda50cc0b7a.pdf, accessed
on 29 March 2017.

Prakash, V. 1999. 'Status of Vultures in Keoladeo National
Park, Bharatpur, Rajasthan, with Special Reference to
Population Crash in *Gyps* Species', *Journal of Bombay
Natural History Society*, 96(3): 365–78.

Prakash, V., R.E. Green, D.J. Pain, S.P. Ranade, S. Saravanan,
N. Prakash, R. Venkitachalam, R. Cuthbert, A.R.
Rahmani, and A.A. Cunningham. 2007. 'Recent Changes
in Populations of Resident *Gyps* Vultures in India', *Journal
of Bombay Natural History Society*, 104: 129–135.

Raghavan, Rajeev, Gopal Prasad, Benno Pereira, P.H. Anvar
Ali, and Lavanchawee Sujarittanonta. 2009. '"Damsel in
Distress"- The Tale of Miss Kerala, *Puntius denisonii* (Day),
An Endemic and Endangered Cyprinid of the Western
Ghats Biodiversity Hotspot (South India)', *Aquatic
Conservation: Marine and Freshwater Ecosystems*, 19(1): 67–74.

Rahmani A.R. 2008. 'Race to Save the Vultures', *Journal of
Bombay Natural History Society*, 148–55.

Rahmani, A.R. 2014. 'Vulture Safety Zones', *Mistnet*,
July–September, 15(3).

Raj, K. Diraviya. 2010. 'Status of Seagrass Diversity,
Distribution and Abundance in Gulf of Mannar Marine
National Park and Palk Bay (Pamban to Thondi),
Southeastern India', *South Indian Coastal and Marine
Bulletin*, 2(2): 1–21.

Select References

Ramesh, K., J. Johnson, S. Sen, R.S. Murthy, M.S. Sarkar, M. Malviya, S. Bhardwaj, M. Naveen, S. Roamin, V.S. Parihar, and S. Gupta. 2013. *Status of Tiger and Prey Species in Panna Tiger Reserve, Madhya Pradesh, India: Capture-Recapture and Distance Sampling Estimates.* Wildlife Institute of India, Dehradun, and Panna Tiger Reserve, Madhya Pradesh. Available at www.researchgate.net/publication/266910362_STATUS_OF_TIGER_AND_PREY_SPECIES_IN_PANNA_TIGER_RESERVE_MADHYA_PRADESH_INDIA, accessed on 29 March 2017.

Ranjitsinh, M.K. 1972. *A Note on the Future Conservation Plan for the Pigmy Hog* (Sus salvanius) *and Hispid Hare* (Caprolagus hispidus *Pearson, 1839*). Unpublished Paper, D.O. No. 682/DSF 1023/72, Govt. of India.

Rhodin, A.G.J., J.B. Iverson, P.P. van Dijk, R.A. Saumure, K.A. Buhlmann, P.C.H. Pritchard, and R.A. Mittermeier. (eds.), 'Turtles of the World: Annotated Checklist and Atlas of Taxonomy, Synonymy, Distribution, and Conservation Status (8th Ed.)', *Conservation Biology of Freshwater Turtles and Tortoises: A Compilation Project of the IUCN/SSC Tortoise and Freshwater Turtle Specialist Group.* Chelonian Research Monographs 7:1–292.

Sahyadri Nisarga Mitra (blog), 'Marine Turtle Conservation in Maharashtra (2002 to 2006)', Available at www.snmcpn.org/turtles/conservation-efforts, accessed on 24 February 2019.

Select References

Shankar, Kartik. 2017. *From Soup to Superstar: The Story of Sea Turtle Conservation along the Indian Coast*. New Delhi: HarperCollins.

Sharma, Sandeep, Trishna Dutta, Jesús Maldonado, Thomas Wood, Hemendra Singh Panwar, and John Seidensticker. 2013. 'Spatial Genetic Analysis Reveals High Connectivity of Tiger (*Panthera tigris*) Populations in the Satpura-Maikal Landscape of Central India', *Ecology and Evolution*, 3(1): 48–60. DOI: 10.1002/ece3.432.

Soulé, Michael, and Reed Noss. 1998. 'Rewilding and Biodiversity: Complementary Goals for Continental Conservation', *Wild Earth*, 8: 18–28.

Tanner, Thomas. 1980. 'Significant Life Experiences: A New Research Area in Environmental Education', *Journal of Environmental Education* 11(4): 20–4. DOI: 10.1080/00958964.1980.9941386.

Wildlife Trust of India, 'Daring to Restore: Coral Reef Recovery in Mithapur', 2008, accessed May 2018.

Acknowledgements

I t takes a village to raise a child. For the birthing of this book, an army of people gave their valuable time, input, and resources, and to them I can only express my heartfelt gratitude.

I have to thank the good people at Oxford University Press, India, for seeing the potential in the book and its philosophy. I'd also like to thank Rekha Natarajan for her encouragement and enthusiasm for the idea. I cannot thank Meenakshi Subramaniam enough for so generously sharing her artwork with us and so promptly sending us the images we needed. (You can check out her work on MeenArt.in). I must also thank Bikram Grewal, as this painting was made at his ornate residence, The Walterre, in Dehradun.

So many scientists went through the manuscript of this book, offering advice from time to time with the necessary facts: Asad Rahmani, Deepak Apte, Vibhu and

Acknowledgements

Nikita Prakash, Adrian Pinder, Rajeev Raghavan, Romulus Whitaker, Shailendra Singh, Raghu Chundawat, Divya Karnad, Vijay Dhasmana, Harini Nagendra, Goutam Narayan, Parag Deka, Rathin Burman, Amit Sharma, and Jeffrey Lang. I received immense institutional support from the Wildlife Trust of India and WWF-India.

I am grateful to Bhavna Menon from the Last Wilderness Foundation and am really impressed with all the wonderful work being done with the Pardhis. I must thank Vivek Talwar, formerly with Tata Chemicals, for being so open to suggestions and for organizing the trip to the mahseer breeding facility in Maharashtra. Mahesh Rangarajan, I thank you for being such a great mentor and inspiring me to write.

On a personal note, my father, who taught us that travel is the best form of education, and that books make better treats than candy. For my daughters, little Aranya and Prakriti: I had to spend a lot of time away from you both while travelling for this book. I hope you will forgive me. May your life be blessed with wondrous moments enjoying nature's splendour. To Vijay, for always keeping me in sharp focus even when I lose sight of the big picture (and, of course, for taking the author picture!)

Finally, this book is dedicated to the people who embrace the spirit of rewilding every day of their lives.

About the Author

Bahar Dutt is one of the foremost voices championing the environment in India. She has worked with CNN-News18 as an environment editor, run a highly successful column for *The Mint* newspaper, and been associated with organizations that focus on conservation and sustainable development, such as the United Nations Environment Programme, the Wildlife Trust of India, and Development Alternatives. Bahar worked on a livelihoods project for snake charmers in rural India and helped set up a rescue centre for injured primates. In 2018, she set up The Mitti Project, for bringing children and adults close to nature. She is the author of the book *Green Wars—Dispatches from a Vanishing World* (2014) that explores the tension between environment and development.

Bahar has won over 12 national and international awards for her stories, including the Ramnath Goenka Award for Environmental Reporting, the Sanskriti Award for community service, the Young Journalist Award from FEJI (Forum for Environment Journalists India), and the Wildscreen Award in Bristol, UK. She has also been featured in *Vogue India* and *Verve* magazines for her work.

An aspiring gardener and a closet baker, Bahar can be found sometimes in the city, and most other times talking to trees in forests. She lives in New Delhi with her husband, Vijay, her daughters, Aranya and Prakriti, and her dog, Musibat.